计算机技术开发与应用丛书

虚拟化KVM极速入门

陈 涛 ◎ 编著

清华大学出版社

北京

内 容 简 介

虚拟化技术是云计算的底层支撑技术之一。作为已经纳入Linux内核的虚拟化解决方案，KVM虚拟化近年来发展迅猛，是很多公共云供应商默认的虚拟机管理程序。对于IT从业者来讲，掌握一些KVM虚拟化知识是很有必要的。

本书是《虚拟化KVM进阶实践》的姊妹篇，共分6章。针对初学者，先从虚拟化基本概念及KVM原理讲起，然后通过全动手的实验学习KVM的安装、虚拟机的创建、虚拟机的管理、虚拟网络的管理、虚拟存储的管理等。

KVM虚拟化其实是Linux、KVM、QEMU和libvirt等很多开源技术的组合，对于初学者来讲学习起来比较难。本书将这些技术融会贯通，是作者多年讲授KVM虚拟化实战课程经验的结晶，可以使学习者快速入门并为后续的企业级应用打下基础。

本书封面贴有清华大学出版社防伪标签，无标签者不得销售。
版权所有，侵权必究。举报: 010-62782989, beiqinquan@tup.tsinghua.edu.cn。

图书在版编目(CIP)数据

虚拟化KVM极速入门/陈涛编著. —北京: 清华大学出版社, 2022.4(2024.9重印)
(计算机技术开发与应用丛书)
ISBN 978-7-302-58987-7

Ⅰ. ①虚… Ⅱ. ①陈… Ⅲ. ①虚拟处理机 Ⅳ. ①TP338

中国版本图书馆CIP数据核字(2021)第173531号

责任编辑: 赵佳霓
封面设计: 吴 刚
责任校对: 时翠兰
责任印制: 宋 林

出版发行: 清华大学出版社
网 址: https://www.tup.com.cn, https://www.wqxuetang.com
地 址: 北京清华大学学研大厦A座　　　　邮 编: 100084
社 总 机: 010-83470000　　　　　　　　　邮 购: 010-62786544
投稿与读者服务: 010-62776969, c-service@tup.tsinghua.edu.cn
质量反馈: 010-62772015, zhiliang@tup.tsinghua.edu.cn
课件下载: https://www.tup.com.cn, 010-83470236

印 装 者: 三河市天利华印刷装订有限公司
经 销: 全国新华书店
开 本: 186mm×240mm　　印 张: 20　　字 数: 451千字
版 次: 2022年4月第1版　　　　　　　　印 次: 2024年9月第4次印刷
印 数: 3801~4800
定 价: 89.00元

产品编号: 089903-01

前言
PREFACE

本书的由来

与 VMware、Microsoft 虚拟化技术相比，KVM 虚拟化对于初学者并不"友好"。作为虚拟化项目的组成部分，笔者从 2011 年开始为客户讲授 KVM 虚拟化的课程，对此感触特别深。为了"不重复发明轮子"，KVM 虚拟化充分利用了 Linux、QEMU 和 libvirt 等开源技术，是一种组合型的解决方案，对初学者要求较高。

根据长期的 KVM 面授课程的经验，笔者总结出这样一种教学方法：针对每个知识点，先学习适当深度的原理，再动手做实验；先通过图形界面的操作，看到大概的轮廓，再通过大量的命令行、脚本的练习强化学习到的知识；先学基本知识，再掌握最佳实践方案。采用这种教学方法，通过 8 天左右的培训，就可以让初学者成为一名合格的 KVM 虚拟化平台的管理员。

2015 年，笔者将面授课程搬到了线上，制作了"开源虚拟化 KVM 入门"和"KVM 虚拟化进阶与提高"两门视频课程，发布在 51CTO 学院上，目前已有约 23 万人参加学习。

随着 RHEL/CentOS 8 的发布，笔者又将这套课程进行更新迭代，形成了《虚拟化 KVM 极速入门》和《虚拟化 KVM 进阶实践》，仍然沿用"原理＋实验"的风格，希望能够帮助读者。

本书内容

本书共分 6 章。通过学习本书，KVM 虚拟化的初学者可以掌握 KVM 虚拟化的原理，以及管理单台宿主机所需要的知识。

第 1 章介绍虚拟化定义与历史，KVM 的原理与架构，KVM 与 QEMU、libvirt 等的关系。

第 2 章介绍如何将一台 Linux 主机配置为虚拟化主机，验证虚拟化功能及管理 KVM 的工具。

第 3 章介绍在 RHEL/CentOS 8 中通过 Cockpit、virt-manager 和 virt-install 等方法创建虚拟机，VirtIO 驱动程序、QEMU Guest Agent 和 SPICE Agent 的工作原理及安装。

第 4 章介绍通过 virt-manager、virsh 和 Cockpit 对虚拟化平台进行日常管理，包括创建、暂停、恢复、停止及删除等生命周期的操作管理。

第 5 章介绍虚拟网络的管理，包括 NAT、桥接、隔离、路由、开放等网络类型的原理与配置，VLAN 和网络过滤器的原理与配置。

第 6 章介绍虚拟存储的管理，包括托管和非托管的存储区别，qemu-img 命令的使用，存储池、存储卷的原理。

如何使用本书

本书既是笔者自己学习和使用 KVM 虚拟化的总结，又是讲授 KVM 虚拟化课程的课件。笔者认为学习原理、动手实践、做好记录、细心排错是学习 KVM 虚拟化的关键。

聪明人下笨功夫。在本书的陪伴下，我们一起：

(1) 深入理解原理。

(2) 精读 man 帮助、官方文档等。

(3) 做所有的实验。

(4) 详细记录实验过程。

(5) 使用思维导图等辅助工具。

(6) 享受排错的过程，在寻求帮助之前先尝试自己解决。

致谢

开源软件的世界精彩万千，在本书的写作过程中参考了很多开源社区的资料。在此向开源社区所有参与者和无私的代码贡献者致敬。

感谢龙芯中科杨昆、田延辉先生对龙芯 CPU 运行 KVM 虚拟机技术细节的介绍。

感谢陈庭暄先生在 CentOS Stream 9 上对全部实验进行的验证工作。

感谢清华大学出版社的工作人员为本书付出的辛勤劳动。

云计算技术发展很快，加之笔者水平有限，书中难免存在疏漏，敬请读者批评指正。

陈　涛

2022 年 1 月

目 录
CONTENTS

本书源代码　　　教学课件（PPT）

第 1 章　KVM 概述 ··· 1

1.1　虚拟化概述 ·· 1
　　1.1.1　虚拟化的定义 ·· 1
　　1.1.2　操作系统虚拟化的历史 ··· 2
　　1.1.3　操作系统虚拟化的实现方式 ····································· 3
　　1.1.4　虚拟化翻译技术的分类 ··· 6
　　1.1.5　Hypervisor 的分类 ·· 11
1.2　KVM 概述 ··· 12
　　1.2.1　KVM 的历史 ··· 13
　　1.2.2　KVM 的体系结构 ··· 14
　　1.2.3　QEMU 与 KVM ··· 14
　　1.2.4　libvirt 与 KVM ·· 15
　　1.2.5　KVM 的集中管理与控制 ·· 15
1.3　本章小结 ··· 17

第 2 章　KVM 安装 ··· 18

2.1　安装环境的准备 ·· 18
　　2.1.1　生产环境的硬件配置 ··· 18
　　2.1.2　实验环境的准备 ··· 18
2.2　KVM 的安装 ·· 23
　　2.2.1　下载 CentOS 8 的 ISO 文件 ····································· 23
　　2.2.2　创建新虚拟机 ·· 24
　　2.2.3　修改虚拟机的设置 ·· 26
　　2.2.4　安装 CentOS 8 时直接安装 KVM 组件 ······················ 30
　　2.2.5　查看安装的结果 ··· 35
　　2.2.6　安装额外组件及升级 ··· 38
　　2.2.7　虚拟化功能验证 ··· 39

2.3 KVM 的管理方法 ·· 41
　　2.3.1 本地管理 ·· 41
　　2.3.2 远程管理 ·· 43
2.4 本章小结 ··· 57

第 3 章 创建虚拟机 ·· 58

3.1 使用 Cockpit 创建虚拟机 ·· 58
　　3.1.1 查看当前配置 ··· 58
　　3.1.2 创建虚拟机 ·· 62
　　3.1.3 查看虚拟机与环境的配置 ······································· 65
3.2 使用 virt-manager 创建虚拟机 ·· 70
　　3.2.1 使用 virt-manager 查看当前配置 ··························· 70
　　3.2.2 创建虚拟机 ·· 73
　　3.2.3 查看虚拟机与环境的配置 ······································· 78
3.3 使用 virt-install 创建虚拟机 ·· 79
　　3.3.1 创建虚拟机并通过交互模式安装 ··························· 80
　　3.3.2 查看虚拟机与环境的配置 ······································· 82
　　3.3.3 virt-install 高级用法示例 ·· 82
3.4 半虚拟化驱动 VirtIO ··· 85
　　3.4.1 半虚拟化驱动 VirtIO 原理 ······································ 85
　　3.4.2 半虚拟化驱动 VirtIO 的安装 ·································· 87
3.5 QEMU Guest Agent ·· 95
　　3.5.1 QEMU Guest Agent 原理 ······································· 95
　　3.5.2 Linux 下的 QEMU Guest Agent ····························· 97
　　3.5.3 Windows 下的 QEMU Guest Agent ······················· 98
3.6 显示设备与协议 ··· 102
　　3.6.1 显示设备 ··· 102
　　3.6.2 显示协议 ··· 103
　　3.6.3 Remote Viewer 连接虚拟机排错 ··························· 107
　　3.6.4 Linux 下的 SPICE Agent ······································· 112
　　3.6.5 Windows 下的 SPICE Agent ································· 113
3.7 本章小结 ··· 114

第 4 章 管理虚拟机 ·· 115

4.1 libvirt 架构概述 ·· 115
4.2 使用 virt-manager 管理虚拟机 ··· 116

        4.2.1　virt-manager 界面概述 ································· 116
        4.2.2　虚拟机生命周期管理 ································· 122
        4.2.3　管理虚拟硬件 ·· 125
    4.3　使用 virsh 管理虚拟机 ·· 130
        4.3.1　获得帮助 ··· 130
        4.3.2　常用的子命令 ·· 132
    4.4　使用 Cockpit 管理虚拟机 ···································· 140
    4.5　本章小结 ·· 140

第 5 章　管理虚拟网络 ·· 141
    5.1　查看默认网络环境 ··· 141
        5.1.1　查看宿主机的网络环境 ································ 141
        5.1.2　查看 libvirt 的网络环境 ······························ 143
        5.1.3　查看虚拟机的网络配置 ································ 147
    5.2　TUN/TAP 设备工作原理与管理 ·································· 149
    5.3　网桥工作原理与管理 ·· 155
        5.3.1　考察现有网桥 ·· 155
        5.3.2　通过 iproute 管理网桥 ······························· 155
        5.3.3　通过 NetworkManager 管理网桥 ······················· 159
        5.3.4　通过网络接口文件管理网桥 ··························· 165
        5.3.5　通过 Cockpit 管理网桥 ······························· 166
    5.4　KVM/libvirt 常用的网络类型 ·································· 168
        5.4.1　虚拟机支持的网络 ···································· 169
        5.4.2　libvirt 管理的虚拟网络 ······························ 172
        5.4.3　NAT 模式 ·· 174
        5.4.4　桥接模式 ·· 176
        5.4.5　隔离模式 ·· 176
        5.4.6　路由模式 ·· 176
        5.4.7　开放模式 ·· 177
        5.4.8　直接附加模式 ·· 177
        5.4.9　PCI 直通与 SR-IOV ··································· 178
    5.5　创建和管理隔离的网络 ··· 179
        5.5.1　通过 virt-manager 创建和管理隔离网络 ··············· 179
        5.5.2　通过 Cockpit 创建和管理隔离网络 ···················· 183
        5.5.3　通过 virsh 创建和管理隔离网络 ······················ 184
        5.5.4　使用隔离网络 ·· 187

5.6 创建和管理 NAT 的网络 ·· 189
   5.6.1 使用多种方式创建 NAT 网络 ·································· 189
   5.6.2 使用 NAT 网络 ·· 193
5.7 创建和管理桥接的网络 ·· 195
   5.7.1 在宿主机上创建网桥 ·· 195
   5.7.2 使用网桥 ·· 198
5.8 创建和管理路由的网络 ·· 200
   5.8.1 在宿主机上创建路由模式的网络 ······························ 200
   5.8.2 使用路由模式的网络 ·· 204
5.9 创建和管理开放的网络 ·· 207
5.10 实现多 VLAN 支持 ··· 212
   5.10.1 创建支持 VLAN 的网络接口 ·································· 212
   5.10.2 创建使用 VLAN 网络接口的网桥 ···························· 218
   5.10.3 配置虚拟机使用 VLAN ······································· 219
5.11 通过网络过滤器提高安全性 ··· 221
   5.11.1 网络过滤器基本原理 ··· 222
   5.11.2 网络过滤器的管理工具 ······································ 226
   5.11.3 预安装的网络过滤器 ··· 227
   5.11.4 网络过滤器语法基本格式 ···································· 230
   5.11.5 自定义网络过滤器示例 ······································ 233
5.12 本章小结 ··· 235

第 6 章　管理虚拟存储 ·· 236

6.1 虚拟存储的术语 ··· 236
   6.1.1 虚拟机的存储设备 ··· 236
   6.1.2 宿主机的存储资源 ··· 239
6.2 非托管的存储 ·· 241
   6.2.1 使用 dd 创建磁盘映像文件 ···································· 241
   6.2.2 使用 virsh 管理虚拟机磁盘映像文件 ························· 244
   6.2.3 使用 virt-manager 管理虚拟机磁盘映像文件 ················ 248
6.3 qemu-img 命令的使用 ·· 251
   6.3.1 qemu-img 支持的映像文件格式 ······························· 252
   6.3.2 创建和格式化新的映像文件 ··································· 253
   6.3.3 检查映像文件的一致性 ·· 254
   6.3.4 重新调整映像文件的大小 ····································· 255
   6.3.5 qcow2 映像文件的选项 ·· 256

6.3.6　基础映像与派生映像 ………………………………………………… 257
  6.3.7　修改映像文件的选项 ………………………………………………… 259
  6.3.8　转换映像文件格式 …………………………………………………… 260
  6.3.9　比较映像文件 ………………………………………………………… 261
  6.3.10　更改基础映像文件 …………………………………………………… 261
  6.3.11　提交对映像文件的更改 ……………………………………………… 263
  6.3.12　显示映像文件布局 …………………………………………………… 263
  6.3.13　快照管理 ……………………………………………………………… 265
6.4　存储池 …………………………………………………………………………… 267
  6.4.1　查看当前存储池 ………………………………………………………… 268
  6.4.2　存储池的分类 …………………………………………………………… 269
  6.4.3　创建存储池的通用流程 ………………………………………………… 270
  6.4.4　基于目录的存储池 ……………………………………………………… 271
  6.4.5　基于物理磁盘的存储池 ………………………………………………… 277
  6.4.6　基于 LVM 卷组的存储池 ……………………………………………… 281
  6.4.7　基于网络文件系统的存储池 …………………………………………… 284
  6.4.8　基于 iSCSI 目标的存储池 ……………………………………………… 290
6.5　存储卷 …………………………………………………………………………… 298
  6.5.1　获得存储卷的信息 ……………………………………………………… 298
  6.5.2　创建存储卷 ……………………………………………………………… 300
  6.5.3　向虚拟机分配存储卷 …………………………………………………… 304
  6.5.4　删除存储卷及擦除存储卷 ……………………………………………… 307
6.6　本章小结 ………………………………………………………………………… 308

第 1 章

KVM 概述

虚拟化是云计算的底层支撑技术之一,目前主要有 ESXi、Xen、Hyper-V、KVM 4 种虚拟化引擎。如今,KVM 可在大多数 Linux 发行版上运行,并且是很多公共云供应商默认的虚拟机管理程序。

本章首先讲解与虚拟化有关的基本知识,然后讲解 KVM 的概念和基本架构。

本章要点:

- 虚拟化概述。
- KVM 概述。

1.1 虚拟化概述

学习 KVM 虚拟化,我们需要掌握一些与虚拟化相关的基本知识,包括虚拟化的概念、发展史和实现方式。

- 虚拟化的定义。
- 操作系统虚拟化的历史。
- 操作系统虚拟化的实现方式。

1.1.1 虚拟化的定义

维基百科对虚拟化的定义是"在计算机技术中,虚拟化(技术)或虚拟技术(Virtualization)是一种资源管理技术,是将计算机的各种实体资源(CPU、内存、磁盘空间、网路适配器等),予以抽象、转换后呈现出来并可供分割、组合为一个或多个计算机组态环境。由此,打破实体结构间的不可切割的障碍,使用户可以比原本的组态以更好的方式来应用这些计算机硬体资源。这些资源的新虚拟部分是不受现有资源的架设方式、地域或物理组态所限制。一般所指的虚拟化资源包括计算能力和资料存储"。

根据应用的对象不同,虚拟化技术分为多种:

(1)操作系统虚拟化:将资源划分为一个或多个执行环境,运行不同的操作系统。

(2)存储虚拟化:对存储资源进行抽象化表现,从早期的 RAID、LVM,到后来 VTL(虚拟磁带库),以及现在的软件定义的存储等都是存储虚拟化。

(3) 网络虚拟化:软件定义的网络(Software Defined Network,SDN)主要分为两个方向:控制平面虚拟化与数据平面虚拟化。

(4) GPU 虚拟化:将图形处理器单元(Graphics Processing Unit,GPU)虚拟化之后,可以让运行在数据中心服务器上的虚拟机实例共享使用同一块或多块 GPU 处理器。

(5) 软件虚拟化:与我们常用的"绿色"软件类似,可以大大降低企业中软件部署、维护、升级的综合成本。

(6) 硬件虚拟化:例如 SR-IOV(Single Root I/O Virtualization)技术对网卡等设备进行封装、管理、共享,在其上创建多个 VF(Virtual Function),每个 VF 对于操作系统来讲都是真实的物理设备。

本书讨论的重点是操作系统虚拟化。

1.1.2　操作系统虚拟化的历史

操作系统虚拟化通常表现为在单一系统上运行多个虚拟操作系统,这些虚拟操作系统同时运行,而且又相互独立。

操作系统虚拟化技术发展史如图 1-1 所示。在 20 世纪 60 年代就已经出现了虚拟化技术,最早是在 IBM CP-40 大型机上使用了虚拟内存和虚拟机的概念。

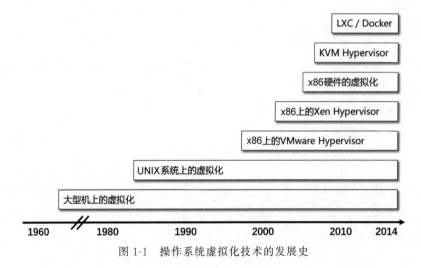

图 1-1　操作系统虚拟化技术的发展史

除此之外,IBM 公司还有很多与虚拟化有关的创新贡献:

(1) 1965 年,推出 System/360 Model 67 和 TSS 分时共享系统(Time Sharing System),允许很多远程用户共享同一高性能计算设备的使用时间。

(2) 1972 年,发布用于创建灵活大型主机的虚拟机(VM)技术,该技术可根据动态的需求快速而有效地使用各种资源。

(3) 1978年，获得了独立磁盘冗余阵列（RAID）概念的专利。
(4) 1991年，为z/OS平台推出分级文件系统（HFS）和网络文件系统（NFS）。
(5) 1997年，推出虚拟磁带服务器（VTS）。该磁带库使用磁盘作为转储磁带数据的高速缓存。
(6) 1999年，在AS/400上第一次推出新的"逻辑分区（LPAR）"技术和新的高可用性集群解决方案。
(7) 2001年，推出基于Power 4处理器的eServer p690，将微处理器技术领域的突破和虚拟化技术相结合。
(8) 2002年，推出Linux虚拟服务（Linux Virtual Services），该服务允许运行Linux应用的客户使用大型主机上的"虚拟服务器"。
(9) 2003年，推出可提供数据块级存储虚拟的SAN卷控制器（SAN Volume Controller），在业界第一次允许客户拥有一个对其存储基础架构进行管理的控制界面。
(10) 2004年，推出了虚拟引擎（Virtualization Engine），大幅度提高了系统和分区的利用率。

在20世纪60—80年代，虚拟化技术在大型机和小型机上获得了空前的成功，并且在相当长的一段时间里，虚拟化技术也只在大型机和小型机上得到了应用。

而后随着x86的流行，整个20世纪80—90年代，基于x86上的虚拟化技术及公司如雨后春笋般地涌现。下面是一些重要的标志性事件：

(1) 1997年，苹果公司开发了Virtual PC，后来将其卖给了Connectix公司。2003年，微软通过收购Connectix公司而获得Virtual PC虚拟化技术。
(2) 1999年，VMware公司针对x86平台推出了VMware Workstation。2001年又发布了ESX和GSX，它们是ESXi的前身。
(3) 2003年，开源虚拟化项目Xen发布，通过半虚拟化技术为x86平台提供虚拟化支持。
(4) 2005—2006年，Intel和AMD相继在x86体系结构的CPU中增加了对虚拟化技术的支持（Intel VT-x，AMD-V）。借助于硬件的力量，虚拟化的性能得到了大大的提升。
(5) 2007年，Linux Kernel 2.6.20合入了由以色列公司Qumranet开发的虚拟化内核模块——基于内核的虚拟机（Kernel-based Virtual Machine，KVM）。
(6) 2008年，LXC（Linux Containers）发布，它可以提供轻量级的虚拟化，以便隔离进程和资源，而且不需要提供指令解释机制及全虚拟化的其他复杂性。

1.1.3 操作系统虚拟化的实现方式

实现操作系统虚拟化主要有3种实现方式：
- 纯软件仿真。
- 虚拟化翻译。
- 容器技术。

下面，我们逐一来了解它们的工作机制。

1. 纯软件仿真

仿真（Emulate）是用一个系统模拟另一个系统。例如某个软件原来是针对系统 A 开发的，它不能直接在系统 B 上运行。假如我们能使系统 B"模拟"系统 A 的工作，那么，这个软件就可以在系统 A 的仿真环境中运行了。

PC 机上的游戏仿真软件就是采用了这个原理。有了这类软件，我们就可以在 PC 玩一些早期的电视游戏机或街机上的游戏，如图 1-2 所示。

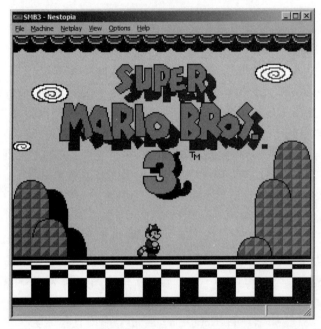

图 1-2　在 Nestopia 模拟器上运行 FC 经典游戏《超级马里奥兄弟》

纯软件仿真实现的虚拟化，顾名思义就是完全通过软件来模拟完整的硬件环境。由于是纯软件的，其优点是可以仿真模拟多种硬件，但缺点是效率比较低。

纯软件仿真类的虚拟化产品主要有 QEMU、Bochs 和 PearPC，其中 QEMU（Quick Emulator 的缩写）最为著名，可以使用纯软件来仿真 x86、MIPS、ARM、PowerPC 等多种 CPU 架构（网站 www.perzl.org 汇集了多个成功案例）。

提示：除了可以实现纯软件仿真以外，QEMU 也可以利用 CPU 的硬件虚拟化特性来提供更好的性能。

---

2. 虚拟化翻译

虚拟化翻译技术看起来与仿真类似，但是其实现机制是完全不一样的。目前多数虚拟化采用了虚拟化翻译技术，这是由虚拟机管理程序 Hypervisor 实现的。Hypervisor 是一个软件层或子系统，也称为虚拟机监控器（Virtual Machine Monitor，VMM）和虚拟器（Virtualizer）。

在学习 Hypervisor 之前，我们需要先了解 2 个在虚拟化中常用的术语：

（1）Host：运行着虚拟机的机器。通常被翻译成主机或宿主机，笔者认为翻译为宿主机更好一些。"宿"本义就是夜晚睡觉，引申有居住、住宿的地方等意思。宿主机就是虚拟机"居住"的地方。

（2）Guest：运行在宿主机上的虚拟机、虚机，也常常被翻译成来宾机。

Hypervisor 可以实现多个 Guest 操作系统在同一个 Host 系统中运行，它为 Guest 操作系统提供虚拟的硬件，如图 1-3 所示。Hypervisor 一直监视 Guest 的运行情况，Guest 对虚拟硬件的访问会被"翻译、转换"成对物理硬件的访问，而不是像纯软件仿真中那样对访问指令进行修改。同时，Hypervisor 还需要控制 Guest 对资源的使用，以及协调处理多个 Guest 同时访问物理资源时发生的冲突，这时 Hypervisor 又充当了"交通警察"的角色。

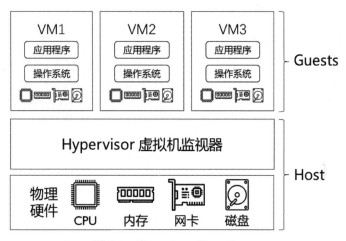

图 1-3　Hypervisor、Host、Guest

纯软件仿真可以"无中生有"地创造出不存在的硬件，而虚拟化翻译却仅限于使用底层已有的硬件，但是，虚拟化翻译最明显的优势就是性能与效率高。

虚拟化翻译技术还有不同的实现方式，在本书的"1.1.4 虚拟化翻译技术的分类"中将会有详细的介绍。

3．容器技术

容器技术最早出现在 Linux 系统中，现在的 Windows 操作系统也很好地支持容器技术。

容器是在 Host 操作系统上构建的一个隔离的轻型接收器，用于在 Host 操作系统上运行应用程序，如图 1-4 所示。容器与容器之间是相互隔离的。容器并不能对 Host 操作系统内核进行自由访问，而只能获取系统的隔离视图。容器与 Host 的用户模式环境隔离。这种轻型隔离环境使应用更易于开发、部署和管理。

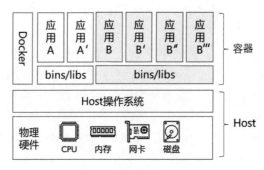

图 1-4 容器体系结构图

容器可以快速启动和停止,因此适用于需要快速适应不断变化的需求的应用,从而大大提高基础结构的密度和利用率。

1.1.4 虚拟化翻译技术的分类

虚拟化翻译技术的核心是 Hypervisor。在讲解 Hypervisor 之前,我们需要了解 CPU 中的保护环(Protection Ring)的基本概念。

CPU 中的保护环是一种用来在发生故障时保护数据、避免恶意操作的方式。工作在不同环中的对象对资源有不同的访问级别,如图 1-5 所示。环是从最高特权级(通常对应最小的数字 0)到最低特权级(最大的数字 3)排列的。在大多数操作系统中,环 0 拥有最高特权,并且可以和最多的硬件直接交互(例如 CPU、内存),同时内层环可以随意使用外层环的资源。

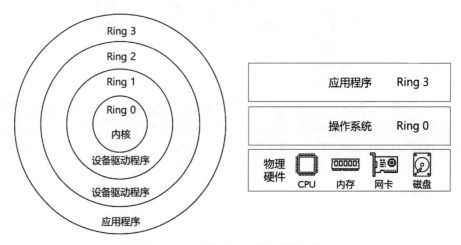

图 1-5 CPU 保护环与传统操作系统

就 x86 CPU 来讲,操作系统内核需要直接访问硬件和内存,因此它的代码需要运行在最高运行级别——环 0 上,这样它就可以执行特权指令、控制中断、修改页表、访问设备等操作。

传统的 x86 操作系统并没有使用环 1 和环 2,操作系统内核和设备驱动程序代码在环 0 中运行,而用户应用程序在环 3 中运行。

应用程序的代码运行在最低运行级别环 3 上,不能直接访问硬件,如图 1-6 所示。如果要想访问硬件,就要通过执行系统调用。执行系统调用的时候,CPU 的运行级别会发生从环 3 到环 0 的切换,并跳转到系统调用对应的内核代码位置执行,这样内核就为应用程序完成了设备访问,完成之后再从环 0 返回环 3。图中的线条①就表示这个系统的调用过程。而线条②表示发生了硬件中断事件。

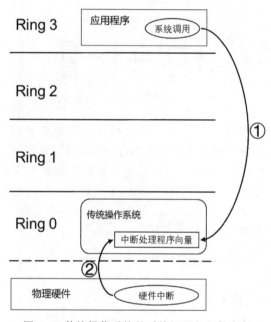

图 1-6 传统操作系统的系统调用与硬件中断

传统的 x86 操作系统原来都是被设计成直接运行在物理硬件设备上的,因此它们会认为自己完全地占有计算机硬件。现在要实现虚拟化,就遇到了一个难题:因为 Host 操作系统已经工作在环 0 上,Guest 操作系统就不能再在环 0 上工作了,但是 Guest 操作系统并不知道这一点,它以前执行什么指令,现在还执行什么指令,但是由于保护环的存在,所以就会出错。这时候,就需要 Hypervisor 来解决这个难题。

根据 Guest 操作系统是如何通过 Hypervisor 实现对硬件的访问,可以将虚拟化翻译技术又细分为 3 种:
- 无硬件辅助的全虚拟化。
- 半虚拟化。
- 有硬件辅助的虚拟化。

1. 无硬件辅助的全虚拟化

无硬件辅助的全虚拟化(Full Virtualization without Hardware Assist)中的硬件主要

是指 CPU,这种虚拟化技术是指不需要特别的 CPU 功能就可以实现的虚拟化。

Hypervisor 运行在环 0 上,它对虚拟机提供虚拟 CPU,如图 1-7 所示。Guest 操作系统运行在环 1 上,不能直接使用真实的物理 CPU,但它并不知道自己是运行在虚拟机之中的。

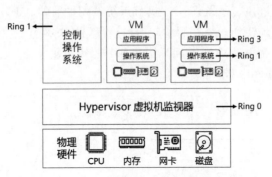

图 1-7 无硬件辅助的全虚拟化

无硬件辅助的全虚拟化是一种基于二进制翻译的全虚拟化(Full Virtualization with Binary Translation),下面分 3 种场景来介绍 Hypervisor 是如何进行翻译转换的。

第一种场景如图 1-8 所示。Guest 操作系统中的应用程序发出系统调用,CPU 捕获到并且给了 Hypervisor 的中断处理程序,Hypervisor 再转给 Guest 操作系统。

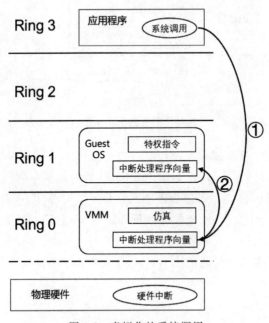

图 1-8 虚拟化的系统调用

第二种场景如图 1-9 所示。当发生硬件中断时,会捕获并给了 Hypervisor 的中断处理程序,然后 Hypervisor 再转给 Guest 操作系统的中断处理程序。

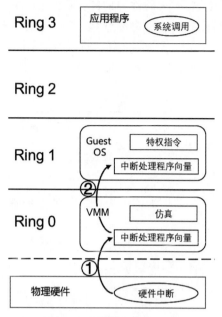

图 1-9　虚拟化的硬件中断

第三种场景如图 1-10 所示。当 Guest 操作系统发出特权指令时,会捕获并给 Hypervisor 进行模拟执行,执行结果返回 Guest 操作系统。

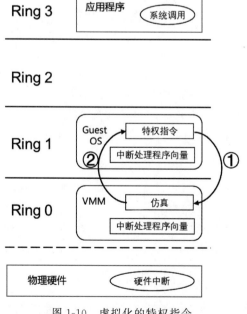

图 1-10　虚拟化的特权指令

从这3种场景可以看出：这种"无硬件辅助的全虚拟化"是一种完全"自力更生"的虚拟化解决方案。它全部靠 Hypervisor 实现翻译转换，很简单的一条指令、一个调用都要通过复杂的处理过程，所以性能损耗大。早期的虚拟化产品都采用这种实现方式，例如 6.0 之前的 VMware Workstation 就是典型的代表产品。为了提高性能，后来就出现了半虚拟化技术。

提示：现在 VMware Workstation 已经到了 Ver16 版本，它可以充分利用 CPU 的虚拟化特性，所以它是硬件辅助的虚拟化解决方案。

2. 半虚拟化

半虚拟化（Para Virtualization）也被称为操作系统辅助的虚拟化，它需要使用修改过后操作系统进行辅助。

Hypervisor 运行在环 0 上。对 Guest 操作系统（通常只有 Linux）内核修改之后，它也可以运行在环 0 上。虚拟机上应用程序还是运行在环 3 上，如图 1-11 所示。

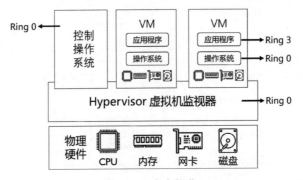

图 1-11 半虚拟化

半虚拟化的思想就是为了减少翻译转换的开销，通过修改 Guest 操作系统的内核代码，通过超级调用（Hyper-call）直接和底层的虚拟化层 Hypervisor 进行通信，如图 1-12 所示。Hypervisor 同时也提供了超级调用接口来满足其他关键内核操作，例如内存管理、中断和时间保持。

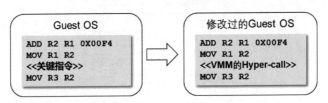

图 1-12 修改 Guest 操作系统

由于这种做法省去了全虚拟化中的捕获和翻译，所以性能损耗非常少，使虚拟机可以接近物理机的性能。这就是半虚拟化的优点，早期 Xen 就凭借此技术大获成功。

但是,半虚拟化技术的缺点也很明显,就是 Guest 操作系统都要有一个专门的定制内核版本,所以,像 Xen 半虚拟化技术只支持虚拟化 Linux,无法虚拟化 Windows。原因很简单:Linux 是开源的,可以修改源代码,而 Windows 操作系统不开源、无法修改源码。

那么,还有什么方法来提高虚拟化的性能呢?这就是下面要讲的"有硬件辅助的虚拟化"。

3. 有硬件辅助的虚拟化

有硬件辅助的虚拟化(Hardware Assisted Virtualization)中的硬件主要指 CPU。这种虚拟化技术可以借助 CPU 中的虚拟化功能,实现更好的性能。

早期的 x86 CPU 是没有虚拟化功能的。随着技术的发展,2005 年 Intel 推出了 Intel VT-x 技术,2006 年 AMD 推出了 AMD-V 技术。这两种技术很相似,如图 1-13 所示,都是将创建一个新的环−1(负 1)给 Hypervisor,这样 Guest 操作系统还可以像之前一样运行在环 0 上。

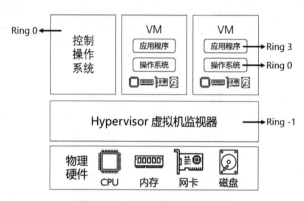

图 1-13　有硬件辅助的虚拟化

对于 Guest 操作系统内核来讲,只要将标志位设为虚拟机状态,就可以直接在 CPU 上执行大部分的指令,这不需要 Hypervisor 在中间翻译转述。除非遇到特别敏感的指令,才需要将标志位设为物理机内核态运行,从而大大提高了效率,这使虚拟化技术更接近物理机的速度。

这是目前应用最广泛的虚拟化技术,VMware ESXi/ESX、Microsoft Hyper-V、Xen 3.0 及之后版本、KVM、VirtualBox、VMware Workstation 6.0 及之后的版本都采用这种虚拟化技术。

1.1.5　Hypervisor 的分类

1974 年,Gerald J. Popek 和 Robert P. Goldberg 的文章 *Formal Requirements for Virtualizable Third Generation Architectures* 将 Hypervisor 分为两种类型。

类型 1:裸金属型

裸金属型(Bare-metal)直接运行在 Host 的硬件上来控制硬件和 Guest 操作系统,如

图 1-14 所示。典型的产品有 VMware ESXi、Microsoft Hyper-V Server、Xen Server。

对于类型 1 的虚拟化来讲，当 Host 加电后，首先加载运行 Hypervisor。从某种意义上讲，也可以将这种 Hypervisor 视为一个特殊的、为虚拟化而优化、裁剪的操作系统。它会完成系统初始化、物理资源的管理等操作系统功能，还负责虚拟机的创建、调度和管理。

有些 Hypervisor 还会提供一个具有特殊权限的虚拟机，它为管理员提供日常管理的操作环境。例如，VMware ESXi 上的 vmKernel 就是一个拥有特殊权限的虚拟机。

类型 2：宿主型

宿主型（Hosted）还是运行在通用操作系统（例如：Windows、Linux）之上，与其他应用程序进程一样，如图 1-15 所示。典型的产品有 VMware Workstation、Windows Server 中的 Hyper-V、Oracle VirtualBox、QEMU、KVM。

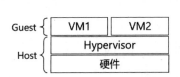
图 1-14　类型 1：裸金属型

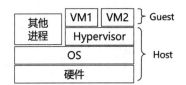
图 1-15　类型 2：宿主型

对于类型 2 的虚拟化来讲，当 Host 加电后，首先运行一个通用的操作系统，我们称之为宿主操作系统。Hypervisor 作为一个特殊的应用程序或内核模块，可以当作宿主操作系统的扩展。Hypervisor 通常不必自己实现对物理资源的管理，所以实现起来比较简洁。

类型 1 与类型 2 之间的主要区别为是否采用了通用操作系统。类型 1 使用专用的 Hypervisor，其优点是专用、体积小和管理成本低，但缺点是可扩展性不强（例如对新硬件的支持不是很及时）。类型 2 的优缺点正好与之相反，其优点是通用的操作系统扩展性强，但缺点是需要更多的管理与维护，不过现在的 Windows Server 和 Linux 发行版本都针对虚拟化应用做了很多优化。

1.2　KVM 概述

KVM 既可以指 KVM 技术本身，也可以指 KVM 开源虚拟化解决方案，如图 1-16 所示。与 VMware、Microsoft 公司的商业虚拟化产品不同，KVM 虚拟化解决方案是由 KVM、QEMU 和 libvirt 这 3 个既独立又合作的开源项目组合在一起而形成的。

图 1-16　KVM 开源虚拟化解决方案

本节将包括以下内容：
- KVM 的历史。
- KVM 的体系结构。
- QEMU 与 KVM。
- libvirt 与 KVM。
- KVM 的集中管理与控制。

1.2.1　KVM 的历史

KVM 是 Kernel Virtual Machine 的缩写，在其官方网站 http://www.Linux-kvm.org 的首页有一个清晰的介绍：

KVM (for Kernel-based Virtual Machine) is a full virtualization solution for Linux on x86 hardware containing virtualization extensions (Intel VT or AMD-V). It consists of a loadable Kernel module, kvm.ko, that provides the core virtualization infrastructure and a processor specific module, kvm-intel.ko or kvm-amd.ko.

Using KVM, one can run multiple virtual machines running unmodified Linux or Windows images. Each virtual machine has private virtualized hardware: a network card, disk, graphics adapter, etc.

KVM is open source software. The Kernel component of KVM is included in mainline Linux, as of 2.6.20. The userspace component of KVM is included in mainline QEMU, as of 1.3.

KVM 是硬件辅助的全虚拟化解决方案。由于它是开源软件，所以除了支持 x86 架构之外，目前已移植到 S/390、PowerPC、IA-64、ARM、MIPS、龙芯等 CPU 中。

KVM 虚拟化最初是由一个以色列的创业公司 Qumranet 开发的。2006 年 10 月，在完成了基本功能、动态迁移及主要的性能优化之后，Qumranet 正式对外宣布了 KVM 的诞生。不到一年时间，KVM 模块的源代码就被正式接纳进 Linux 内核，成为内核的一个模块（在 2007 年 2 月发布的 Linux 内核 2.6.20 中正式包含了 KVM）。

在当时，作为一个功能和成熟度都逊于 Xen 的开源项目，在这么短的时间被内核社区所接纳，其中还有一个故事：当时虚拟化技术方兴未艾，内核社区也急于将对虚拟化的支持包含在内。Xen 的做法是自己而不由 Linux 内核来管理系统资源，这种架构引起了内核开发人员的不满和抵触，所以 Linux 内核最终与 KVM 结缘。

结缘之后，KVM 得到了飞速发展。2008 年 9 月，Red Hat 公司出资收购了 Qumranet 公司。这样，Red Hat 公司就有了自己的虚拟化解决方案，于是开始在自己的产品中用 KVM 替换 Xen（2010 年 11 月发行的 RHEL6 将 Xen 替换为 KVM）。

KVM 作为可加载的内核模块包含在 Linux 内核之中，除了通用模块 kvm.ko 之外，针对不同的 CPU 类型，还需要有特定的模块，例如针对 Intel 的 kvm-intel.ko，针对 AMD 的 kvm-amd.ko。加载这些模块之后，一台 Linux 主机摇身一变便成为 Hypervisor，就可以运行虚拟机了。当然,启用 KVM 内核模块是可选的,例如一台仅仅用于运行数据库服务器软件的 Linux 主机,就没有必要启用 KVM 内核模块。

1.2.2 KVM 的体系结构

KVM 的体系结构的核心是一组实现虚拟化功能的 Linux 内核模块,如图 1-17 所示。这包括提供核心的虚拟化能力的可加载的模块 kvm.ko,还包括 kvm-intel.ko、kvm-amd.ko 等与特定 CPU 相关的模块。它们的主要功能是初始化 CPU 硬件,打开虚拟化模式以支持虚拟机的运行。

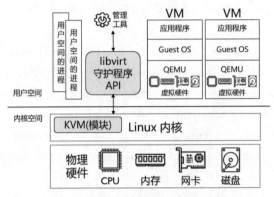

图 1-17 KVM 解决方案体系结构

以 Intel 的 CPU 为例,在被内核加载后,KVM 会先初始化内部数据结构,检查系统当前 CPU,然后打开 CPU 控制寄存器 CR4 中的虚拟化模式开关,并通过 VMXON 指令将宿主机操作系统置于虚拟化模式中的 root 模式,还会创建一个特殊的设备文件 /dev/kvm 并等待用户空间的命令。可以通过 file 命令来查看此文件:

```
#file /dev/kvm
/dev/kvm: character special (10/232)
```

KVM 负责 CPU 和内存的虚拟化,同时也控制着对性能要求比较高的设备,如中断控制器、时钟等设备,而其他的设备例如网卡、显卡、存储控制器和磁盘,都交由 QEMU 来负责模拟。QEMU 是用户空间的程序,它通过 KVM 创建的 /dev/kvm 接口设置一个 Guest 操作系统的地址空间,向它提供模拟的 I/O 设备。

虚拟机的创建和运行是由 QEMU 和 KVM 模块协同来完成的。一个虚拟机就是一个 Linux 进程,通过对这个进程的操作,就可以完成对虚拟机的管理。

1.2.3 QEMU 与 KVM

如前所述,为了启动虚拟机,除了要加载 KVM 内核模块之外还需要有虚拟硬件设备的支持,这将由 QEMU 实现。

在"1.1.3 操作系统虚拟化的实现方式"中我们介绍过QEMU（Quick Emulator）。它是一个开源的仿真器项目，可以通过软件实现多种虚拟硬件。为了提高性能，KVM的开发者对QEMU进行优化及改进从而形成了qemu-kvm。

qemu-kvm通过调用/dev/kvm接口，将与CPU指令有关的部分交由KVM来处理，如图1-18中的①所示。这样qemu-kvm可以直接和安全地将虚拟机的指令直接发送给CPU而无须进行转换。

qemu-kvm还可以模拟出网卡、显卡、存储控制器和硬盘等虚拟设备。因为这还需要由用户空间的驱动、工具实现，所以这些虚拟设备的性能会比较差，如图1-18中的②所示。

为了解决这一问题，还可以通过使用支持透传（Pass Through）的半虚拟化设备来提高存储和网络的性能，例如 virtio_blk、virtio_net，如图1-18中的③所示。

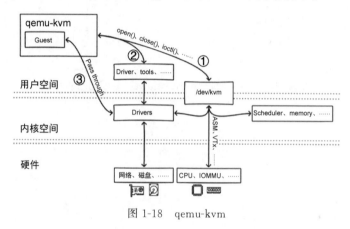

图1-18　qemu-kvm

1.2.4　libvirt 与 KVM

libvirt是一个管理虚拟机和其他虚拟化功能的软件集合，包括API库、守护进程（libvirtd）和其他一些工具，它们在云计算的解决方案中得到很广泛的应用。

在KVM开源虚拟化解决方案中，libvirt扮演着管理工具的角色。用户空间中丰富多彩的管理工具与libvirt进行交互。除了KVM以外，它现在还可以管理Xen、VMware、Hyper-V等虚拟化系统，如图1-19所示。

1.2.5　KVM 的集中管理与控制

在生产环境中，通常会有多台KVM主机，这就需要对它们进行集中化管理与控制。

在KVM官方网站上 http://www.Linux-kvm.org/page/Management_Tools 给出了一个管理软件的列表，其中oVirt项目最为出名。oVirt是一个开源分布式虚拟化解决方案，是Red Hat Virtualization（前身是Red Hat Enterprise Virtualization，RHEV）的上游版本。它可以管理整个企业虚拟化基础架构。

oVirt架构如图1-20所示，主要包含以下3个组件。

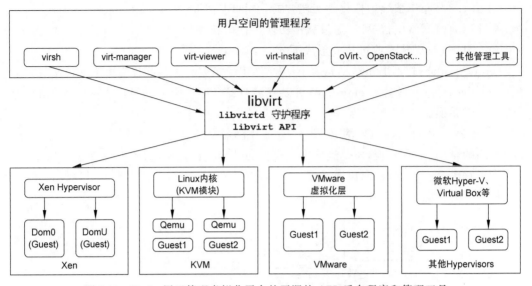

图 1-19 libvirt 用于管理虚拟化平台的开源的 API、后台程序和管理工具

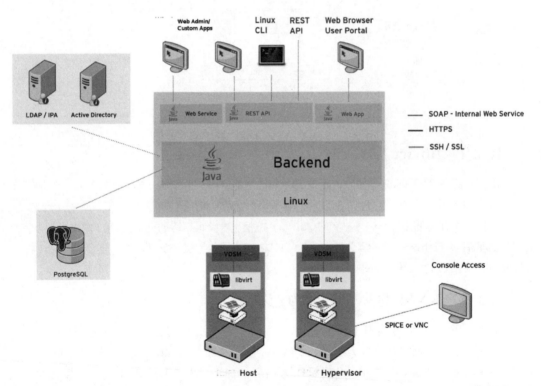

图 1-20 oVirt 架构（来自 oVirt 官方网站）

(1) oVirt 引擎：用于部署、监视、移动、停止和创建 VM 映像，配置存储、网络等。
(2) 计算节点：运行虚拟机的宿主机。
(3) 存储节点：包含与虚拟机的映像和 ISO 文件。

在 RHEL/CentOS 8 中，Cockpit 是默认的服务器管理工具。它是一个基于 Web 的 GUI 管理工具，管理员可以通过该工具监控和管理 Linux 服务器。通过添加插件，Cockpit 还可以管理多台服务器、容器、虚拟机中的网络和存储，以及检查系统和应用的日志，如图 1-21 所示。

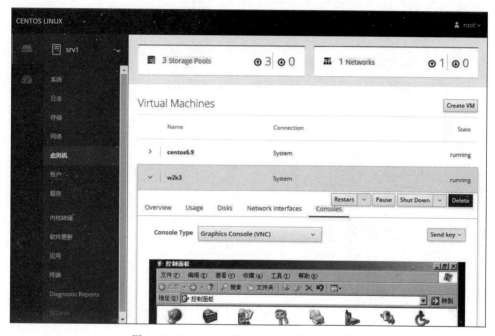

图 1-21　CentOS 8 下 Cockpit 中的虚拟机管理

本书后续的章节将详细介绍 Cockpit 和 oVirt。

1.3　本章小结

本章介绍了虚拟化的定义与历史，以及实现虚拟化的不同方法。知道了 KVM 开源虚拟化解决方案是由 KVM、QEMU 和 libvirt 等组成的。

第 2 章将讲解如何安装配置 KVM 虚拟化主机及 libvirt 管理工具。

第 2 章 KVM 安装

第 1 章介绍了虚拟化的基本概念和 KVM 的原理与架构。本章将讲解如何将一台 Linux 主机配置为虚拟化主机,即通过 KVM 实现虚拟化并且将 libvirt 作为虚拟化管理引擎。

本章要点:
- 安装环境的准备。
- KVM 的安装。
- KVM 的管理方法。

2.1 安装环境的准备

根据应用场景的不同,安装环境的准备分为两种:
- 生产环境的硬件配置。
- 实验环境的准备。

2.1.1 生产环境的硬件配置

在生产环境中,通常使用标准的 x86 服务器作为虚拟化主机。与其他业务不同,虚拟化主机要求服务器上的 CPU 必须支持虚拟化技术,而且必须在 BIOS 中启用对虚拟化技术的支持。

以 Dell PowerEdge R730 服务器为例,它的 BIOS 的设置如图 2-1 所示。在 Processor Setting 中,需要将 Virtualization Technology 选项设置为 Enabled。

其他硬件,如 RAID 卡的配置,与传统的应用基本相同。这里就不赘述了。

2.1.2 实验环境的准备

作为 IT 从业人员,拥有一个属于自己的实验环境是很重要的。构建 KVM 的学习实验环境最便捷的方法就是采用嵌套虚拟化。

嵌套虚拟化类似于俄罗斯套娃娃,是在一个虚拟机之中再运行另外一个虚拟机。嵌套虚拟化是先在第一层虚拟化软件中创建一个虚拟机,然后在此虚拟机上再启用虚拟化功能、安装第二层虚拟化软件,如图 2-2 所示。

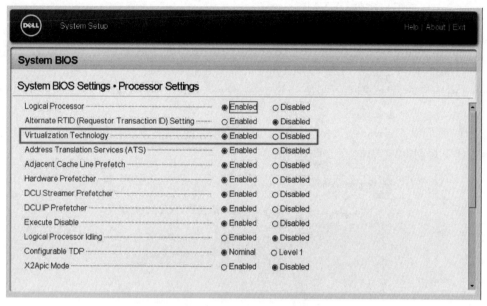

图 2-1　Dell PowerEdge R730 服务器的 BIOS 设置

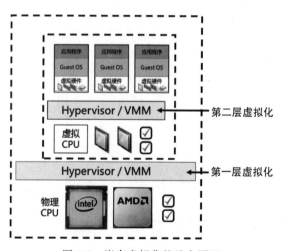

图 2-2　嵌套虚拟化的基本原理

嵌套虚拟化的关键点是:修改"父"Hypervisor 的设置,将其 CPU 的虚拟化特性传递给"子"Hypervisor。目前比较常用的嵌套虚拟化有两种形式:

1. VMware 嵌套虚拟化

(1) 第一层虚拟化软件:VMware Workstation Pro 或 Player、VMware ESXi。

(2)第二层虚拟机软件：KVM、VMware ESXi、Microsoft Hyper-V、Xen。

2．KVM 嵌套虚拟化

(1)第一层虚拟化软件：KVM。

(2)第二层虚拟机软件：KVM、VMware ESXi、Microsoft Hyper-V、Xen。

VMware Workstation Player 是一款供个人免费使用的桌面虚拟化应用程序。在后续的实验中，我们将通过它来构建一个低成本的嵌套虚拟化实验环境。VMware Workstation Player 的详细介绍及下载网址可参见 https://www.vmware.com/cn/products/workstation-player.html。

下面，我们以 VMware-player-16.1.0-17198959.exe 为例来学习它的安装。安装过程可参照图 2-3 至图 2-7 的提示步骤即可。

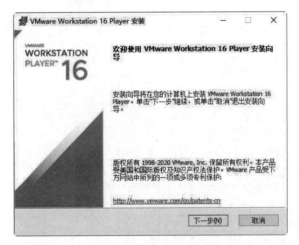

图 2-3　安装向导

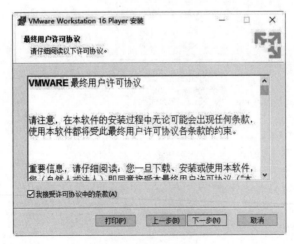

图 2-4　许可协议

第2章 KVM安装 21

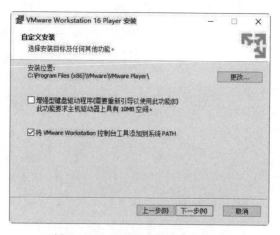

图 2-5 选择安装目标及其他功能

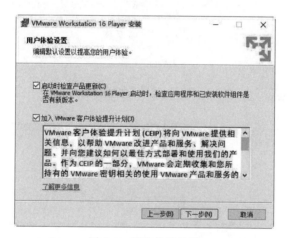

图 2-6 用户体验设置

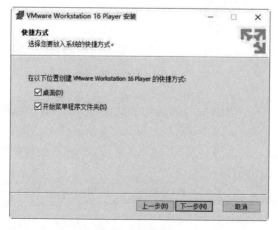

图 2-7 快捷方式

单击"安装"按钮准备开始安装,如图 2-8 所示。

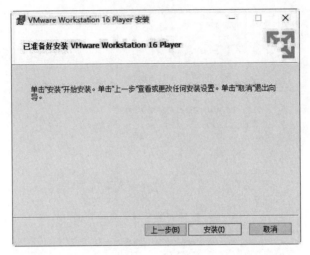

图 2-8　准备安装

单击"完成"按钮完成安装,如图 2-9 所示。

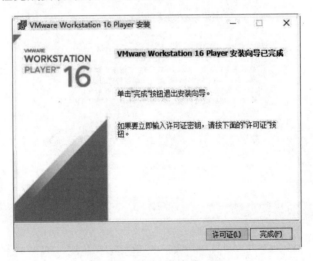

图 2-9　安装完成

安装完成后,我们查看计算机的网络属性的变化,这对以后理解虚拟网络会有帮助。系统新增 2 个网络连接 VMware Network Adapter VMnet1 和 VMware Network Adapter VMnet8,如图 2-10 所示。它们分别与 VMware Workstation 软件中的仅主机(Host-Only)模式和 NAT 模式的虚拟机相连接。

再查看系统服务,系统新增 4 个与 VMware Workstation 软件相关的服务,如图 2-11 所示。

图 2-10　新增加的网络连接

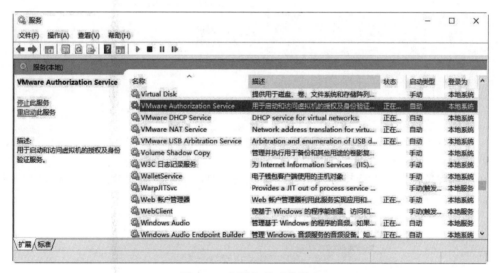

图 2-11　新增加的系统服务

2.2　KVM 的安装

2.2.1　下载 CentOS 8 的 ISO 文件

在后续的实验中,我们将使用 CentOS 8 来构建虚拟化宿主机,所以需要从 CentOS 官方网站(https://www.centos.org/)下载 ISO 文件。例如:CentOS-8.2.2004-x86_64-dvd1.iso,文件大小约为 7.66GB。

2.2.2 创建新虚拟机

在 VMware Workstation Player 主界面中单击"创建新的虚拟机",或者从菜单中选择"Player→文件→新建虚拟机",就可以启动新建虚拟机向导,如图 2-12 所示。

图 2-12　VMware Workstation Player 主界面

在新建虚拟机向导中选择客户机操作系统的来源,选中"稍后安装操作系统",便会创建一个具有空白硬盘的虚拟机,如图 2-13 所示。

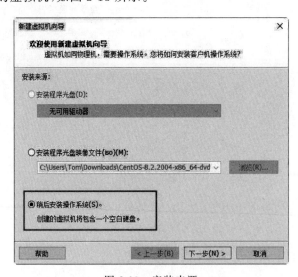

图 2-13　安装来源

接下来,需要指定客户机操作系统的种类与版本。将客户机操作系统类型指定为 Linux,并选择版本为"CentOS 8 64 位",如图 2-14 所示。

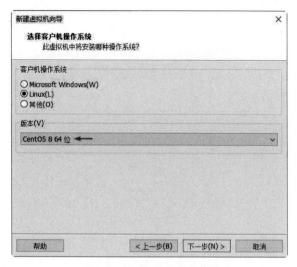

图 2-14　客户机操作系统类型与版本

将虚拟机默认名称修改为更有意义的名称,例如 KVM1,然后将虚拟机文件所在的目录指定到适当的位置,例如固态硬盘而不是普通机械磁盘,这可以大大提高使用的体验,如图 2-15 所示。

图 2-15　虚拟机名称和位置

建议将虚拟机的磁盘设置得大一些,这样以后做实验会更方便一些,例如从 20GB 修改为 80GB。同时选中"将虚拟磁盘存储为单个文件",这样就不会拆分虚拟磁盘文件了,从而可以获得更好的性能,如图 2-16 所示。

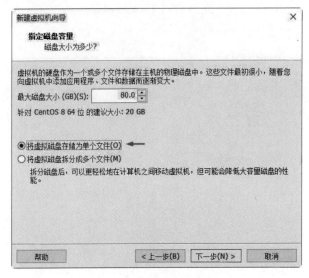

图 2-16　虚拟磁盘设置

单击"完成"按钮以创建虚拟机,如图 2-17 所示。

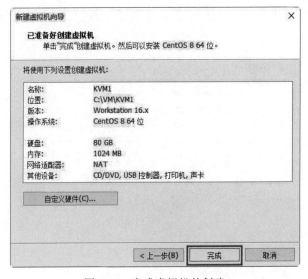

图 2-17　完成虚拟机的创建

2.2.3　修改虚拟机的设置

对于 KVM 虚拟化实验来讲,通过向导创建的虚拟机的配置并不是最优的,例如内存、处理器内核数量少,未启用 CPU 虚拟化特性,还有一些用不着的虚拟硬件,所以需要进行调整。单击"编辑虚拟机设置"打开虚拟机设置,如图 2-18 所示。

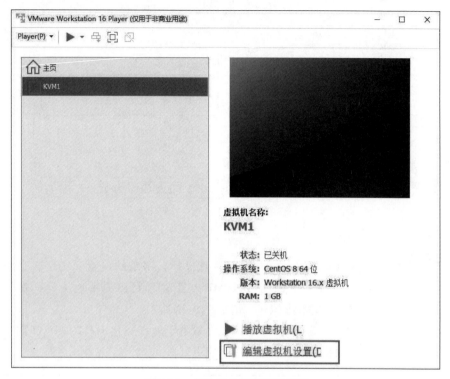

图 2-18　虚拟机的简要信息页面

根据物理机具体硬件条件来修改虚拟机的内存和 CPU 数量,如图 2-19(a)所示,将内存调整为 4GB 内存。

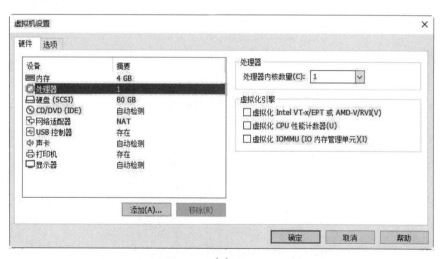

(a)

图 2-19　处理器设置

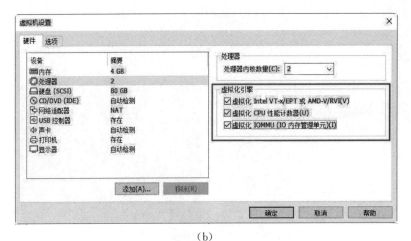

(b)

图 2-19 (续)

除了需要调整处理器内核数量之外,还要注意虚拟化引擎的设置。实现嵌套虚拟化的关键是要选中"虚拟化 Intel VT-x/EPT 或 AMD-V/RVI",可以简单地认为这是将物理机 CPU 的虚拟化特性"传递"给这台虚拟机,如图 2-19(b)所示。

建议将另外 2 个特性即"虚拟化 CPU 性能计数器"和"虚拟化 IOMMU(IO 内存管理单元)"也选中,这样嵌套虚拟化的特性更丰富、性能更好。

注意:在有些型号的 CPU 中,如果虚拟机设置并选中了"虚拟化 CPU 性能计数器",在启动虚拟机时会报错。如果遇到这种情况,则在虚拟机设置中取消这个特性即可。

在 CD/DVD 设置中,选中"启动时连接"。选中"使用 ISO 映像文件",然后单击"浏览"按钮,打开文件选择对话窗口,如图 2-20 所示。

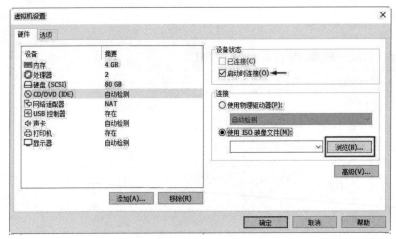

图 2-20 CD/DVD 设置

在文件选择对话窗口中指定前面已经下载好的 CentOS 的 ISO 文件,如图 2-21 所示。

图 2-21 浏览 ISO 映像文件

此时便会在"浏览"按钮左侧显示被指定的文件目录及文件,如图 2-22 所示。

图 2-22 指定 ISO 文件

为了提高虚拟机的整体性能,建议移除 USB 控制器、声卡、打印机这 3 个在后续实验中用不到的虚拟设备,如图 2-23 所示。

再次检查一下硬件配置,最后单击"确定"按钮保存修改,如图 2-24 所示。

注意:在虚拟机网络连接中,使用 NAT 连接模式会比较方便一些。这种模式既不会对现在的网络环境产生影响,虚拟机又可以访问外网。

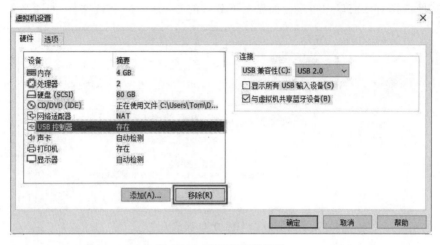

图 2-23 USB 控制器设置

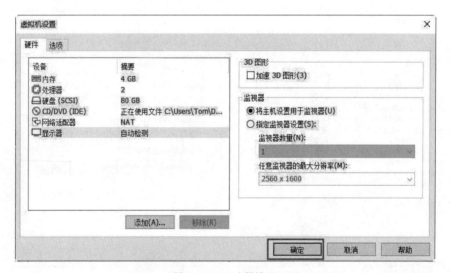

图 2-24 显示器设置

2.2.4 安装 CentOS 8 时直接安装 KVM 组件

现在我们拥有了一个没有操作系统的"空白"虚拟机,下面就需要启动虚拟机,通过从 ISO 文件模拟出的光盘来安装 CentOS 8。在安装过程中,主要配置这些选项:

(1)安装目标。
(2)软件包选择。
(3)时区与日期时间。
(4)网络连接。
(5)root 用户密码。

安装程序的主界面显示有 LOCALIZATION、SOFTWARE、SYSTEM 三类设置，如图 2-25(a)所示。单击 Installation Destination 会出现如图 2-25(b)所示的窗口。仅需要保持默认的磁盘布置，单击 Done 按钮返回安装的主界面。

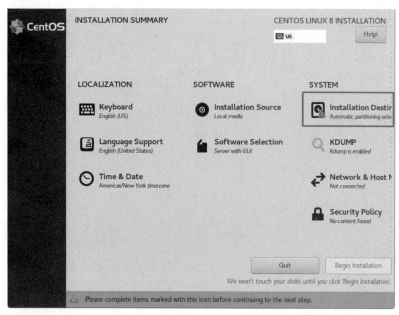

(a)

(b)

图 2-25　安装界面(一)

在安装程序的主界面中单击 Software Selection,如图 2-26(a)所示,就会出现软件选择窗口,如图 2-26(b)所示。

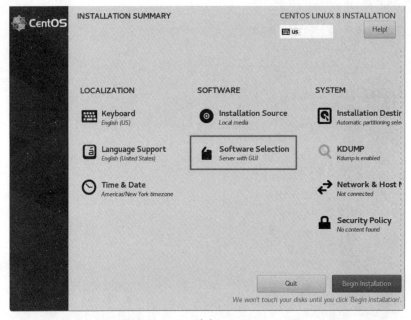

(a)

(b)

图 2-26　安装界面(二)

在左边窗格中选中 Virtualization Host，然后在右边窗格中选中 Remote Management for Linux 和 Virtualization Platform 这 2 个额外的组件，然后单击 Done 按钮返回安装的主界面。

在安装程序的主界面中单击 Network & Host Name，如图 2-27(a)所示，就会出现网络与主机名设置窗口，如图 2-27(b)所示。单击 ON 按钮激活网络连接。由于虚拟机的网络采用了 NAT 模式，所以这个虚拟网卡会从系统的 VMware DHCP Service 服务获得一个动态的 IP 地址配置。将主机名修改为有含义的名称，然后单击 Done 按钮返回安装的主界面。

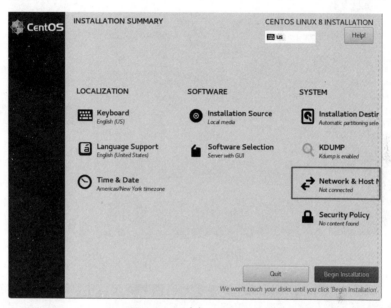

(a)

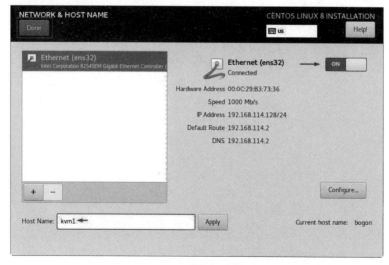

(b)

图 2-27　安装界面(三)

当主要选项设置就绪之后,就可以单击 Begin Installation 按钮进行安装了,如图 2-28 所示。在安装的过程中,建议为 root 用户设置密码,如图 2-29 所示。

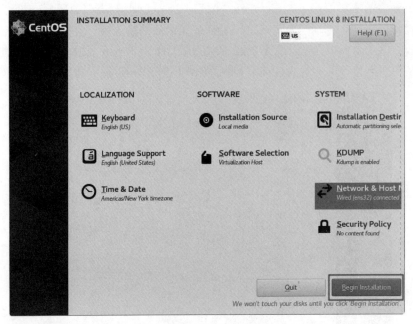

图 2-28　开始安装

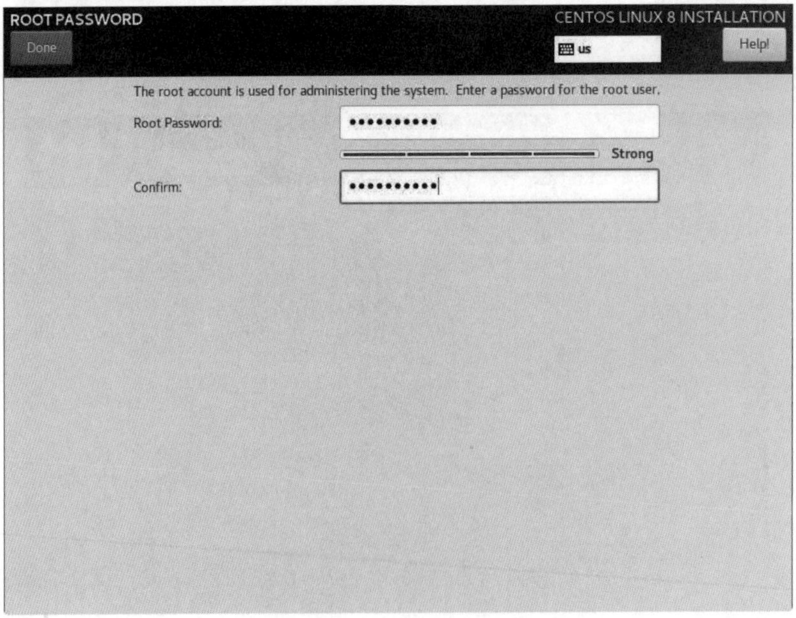

图 2-29　设置 root 密码

安装完成后单击 Reboot 按钮来重新启动系统以完成安装,如图 2-30 所示。

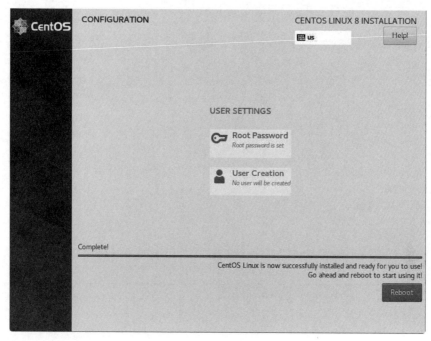

图 2-30　安装结束

2.2.5　查看安装的结果

安装完成后,通过查看安装结果来理解 KVM 的安装。

首先,查看安装程序都安装了哪些软件包。我们可以通过查看 root 用户的主目录中的 anaconda-ks.cfg 文件来查看已经安装的软件包组和软件包,示例命令如下:

```
# cat ~/anaconda-ks.cfg
...
%packages
@^virtualization-host-environment
@remote-system-management
@virtualization-platform
kexec-tools

%end
...
```

其中 virtualization-host-environment 对应的就是图 2-26(b)安装软件选中的 Virtualization Host,@remote-system-management 对应的是 Remote Management for Linux,@virtualization-platform 对应的是 Virtualization Platform。

然后,查看这 3 个软件包组,示例命令如下:

```
# dnf group info virtualization-host-environment
  Environment Group: Virtualization Host
   Description: Minimal virtualization host.
   Mandatory Groups:
     Base
     Core
     Standard
     Virtualization Hypervisor
     Virtualization Tools
   Optional Groups:
     Debugging Tools
     Network File System Client
     Remote Management for Linux
     Virtualization Platform
```

提示:RHEL/CentOS 8 中推荐使用新的包管理器 DNF(Dandified Yum)。它将替代原来的包管理器 YUM。DNF 与 YUM 的命令与参数是兼容的。

环境软件包组 virtualization-host-environment(Virtualization Host)是虚拟化宿主机的最小软件环境。在生产中的专用虚拟化计算节点只需安装这个软件包组。它包含了以下 5 个必需的软件包。

(1) Core:最小的 Linux 环境。

(2) Base:Linux 的基本安装。

(3) Standard:Linux 的标准安装。

(4) Virtualization Hypervisor:最小的虚拟化宿主机环境,仅包括 libvirt 和 qemu-kvm。

(5) Virtualization Tools:可离线管理虚拟映像的工具,主要包括 libguestfs 和 virtio-win。

软件包组 remote-system-management(Remote Management for Linux)中有我们所需要的网页版图像化服务管理工具 Cockpit。它可以管理包括虚拟化在内的多种服务。通过 dnf 命令可查看其信息,示例命令如下:

```
# dnf group info remote-system-management
  Group: Remote Management for Linux
   Description: Remote management interface for CentOS Linux.
   Default Packages:
     cockpit
     net-snmp
     net-snmp-utils
     openwsman-client
```

```
        sblim-cmpi-base
        tog-pegasus
        wsmancli
     Optional Packages:
        openwsman-server
        sblim-indication_helper
        sblim-sfcb
        sblim-wbemcli
```

软件包组 virtualization-platform(Virtualization Platform)中提供了用于访问和控制虚拟机和容器的接口。通过 dnf 命令查看其信息,示例命令如下:

```
# dnf group info virtualization-platform

  Group: Virtualization Platform
  Description: Provides an interface for accessing and controlling virtualized guests and
containers.
     Mandatory Packages:
        libvirt
        libvirt-client
        virt-who
     Optional Packages:
        fence-virtd-libvirt
        fence-virtd-multicast
        fence-virtd-serial
        perl-Sys-Virt
```

提示:根据前面查看的结果,我们就知道:如果在安装 CentOS 8 的时候没有选择虚拟化组件,则还可以通过命令 dnf group install virtualization-hypervisor virtualization-platform virtualization-tools 来添加虚拟化组件。

CentOS 的安装程序还会启动一个名为 libvirtd 的守护程序,它是虚拟化管理系统的服务器端守护程序。libvirtd 可以为虚拟化客户机(如:virsh 命令)执行所需的管理任务,例如:虚拟机的启动、停止和迁移,以及配置网络、存储等资源。通过 systemctl 命令查看状态,示例命令如下:

```
# systemctl status libvirtd
   ● libvirtd.service - Virtualization daemon
     Loaded: loaded (/usr/lib/systemd/system/libvirtd.service; enabled; vendor preset:
enabled)
     Active: active (running) since Thu 2020-07-09 20:09:22 CST; 49min ago
       Docs: man:libvirtd(8)
             https://libvirt.org
```

```
    Main PID: 1225 (libvirtd)
      Tasks: 19 (limit: 32768)
     Memory: 41.7M
     CGroup: /system.slice/libvirtd.service
             ├─1225 /usr/sbin/libvirtd
             ├─1470 /usr/sbin/dnsmasq --conf-file=/var/lib/libvirt/dnsmasq/default.
conf --leasefile-ro --dhcp-script=/usr/libexec/l>
             └─1471 /usr/sbin/dnsmasq --conf-file=/var/lib/libvirt/dnsmasq/default.
conf --leasefile-ro --dhcp-script=/usr/libexec/l>

  Jul 09 20:09:23 kvm1 dnsmasq[1465]: listening on virbr0(#3): 192.168.122.1
  Jul 09 20:09:23 kvm1 dnsmasq[1470]: started, version 2.79 cachesize 150
  …
```

另外,宿主机上还增加了一个名为 virbr0 的网桥(可以理解为一台虚拟交换机)、一个名为 virbr0-nic 的 tun 设备(一种虚拟网络设备)。通过 nmcli 命令查看设备列表,示例命令如下:

```
# nmcli device
DEVICE        TYPE       STATE        CONNECTION
ens32         ethernet   connected    ens32
virbr0        bridge     connected    virbr0
lo            loopback   unmanaged    --
virbr0-nic    tun        unmanaged    --
```

2.2.6 安装额外组件及升级

为了安装和管理虚拟机,建议再添加 Virtualization Client 软件包组,示例命令如下:

```
# dnf group info virtualization-client
  Group: Virtualization Client
  Description: Clients for installing and managing virtualization instances.
  Mandatory Packages:
    gnome-boxes
    virt-install
    virt-manager
    virt-viewer
  Default Packages:
    virt-top
  Optional Packages:
    libguestfs-inspect-icons
    libguestfs-tools
    libguestfs-tools-c

# dnf group install virtualization-client
```

这个软件包组中的 virt-manager、virt-viewer 是两个图形界面的管理工具。如果需要在虚拟化宿主机上运行这两个工具，则还需要安装图形桌面环境，示例命令如下：

```
# dnf group info gnome-desktop
  Group: GNOME
  Description: GNOME is a highly intuitive and user-friendly desktop environment.
  Mandatory Packages:
    ModemManager
    NetworkManager-adsl
    PackageKit-command-not-found
    PackageKit-gtk3-module
    at-spi2-atk
    ...

# dnf group install gnome-desktop
```

最后，建议进行一下联机升级，然后重新启动一下系统，示例命令如下：

```
# dnf -y update

# reboot
```

2.2.7 虚拟化功能验证

最后，我们还需要检查虚拟化功能是否正常。最简单的方法就是检查 CPU 的特性，示例命令如下：

```
# egrep '(vmx|svm)' /proc/cpuinfo
  flags : fpu vme de pse tsc msr pae mce cx8 apic sep mtrr pge mca cmov pat pse36 clflush mmx fxsr
sse sse2 ss syscall nx pdpe1gb rdtscp lm constant_tsc arch_perfmon nopl xtopology tsc_reliable
nonstop_tsc cpuid pni pclmulqdq vmx ssse3 fma cx16 pcid sse4_1 sse4_2 x2apic movbe popcnt tsc_
deadline_timer aes xsave avx f16c rdrand hypervisor lahf_lm abm 3dnowprefetch cpuid_fault
invpcid_single pti ssbd ibrs ibpb stibp tpr_shadow
  ...
```

其中，vmx 是 Intel 的虚拟化功能的标记，svm 是 AMD 虚拟化功能的标记。如果 egrep 命令有输出，则说明这台主机的虚拟化的核心功能是正常的。

另外，在 libvirt-client 软件包中有一个名为 virt-host-validate 的命令，它可以更加全面地验证当前主机的虚拟化功能，示例命令如下：

```
# whatis virt-host-validate
  virt-host-validate (1) - validate host virtualization setup

# virt-host-validate
```

```
QEMU: Checking for hardware virtualization                          : PASS
QEMU: Checking if device /dev/kvm exists                            : PASS
QEMU: Checking if device /dev/kvm is accessible                     : PASS
QEMU: Checking if device /dev/vhost-net exists                      : PASS
QEMU: Checking if device /dev/net/tun exists                        : PASS
QEMU: Checking for cgroup 'cpu' controller support                  : PASS
QEMU: Checking for cgroup 'cpuacct' controller support              : PASS
QEMU: Checking for cgroup 'cpuset' controller support               : PASS
QEMU: Checking for cgroup 'memory' controller support               : PASS
QEMU: Checking for cgroup 'devices' controller support              : PASS
QEMU: Checking for cgroup 'blkio' controller support                : PASS
QEMU: Checking for device assignment IOMMU support                  : PASS
QEMU: Checking if IOMMU is enabled by Kernel                        : WARN (IOMMU appears
to be disabled in Kernel. Add intel_iommu=on to Kernel cmdline arguments)
```

我们注意到这台宿主机上有一个与 IOMMU 有关的警告。在宿主机上启用 IOMMU 功能可以直接将硬件资源分配给虚拟机，这样可以提升虚拟机的性能，但默认情况下 Linux 内核中并未启用 IOMMU 功能。

提示：IOMMU（Input/Output Memory Management Unit）是一个内存管理单元（Memory Management Unit），它可以把设备访问的虚拟地址转化成物理地址。在 KVM 虚拟化中，可以启用 PCI 直通（PCI passthrough）从而允许虚拟机使用主机设备，就像该设备直接连接到虚拟机一样。要启用 PCI 直通功能，就需要启用虚拟化扩展和 IOMMU 功能。

下面，我们就通过修改内核参数来开启这一功能。

首先编辑 grub 配置文件，找到 GRUB_CMDLINE_LINUX。如果主机是 Intel 的 CPU，则就在最后添加 intel_iommu=on。如果主机是 AMD 的 CPU，则就在最后添加 amd_iommu=on，示例命令如下：

```
# vi /etc/default/grub
...
GRUB_CMDLINE_LINUX="crashKernel=auto resume=/dev/mapper/cl-swap rd.lvm.lv=cl/root rd.lvm.lv=cl/swap rhgb quiet intel_iommu=on"
...
```

然后刷新 grub.cfg 文件并重新引导主机，以使这些更改生效，示例命令如下：

```
# whatis grub2-mkconfig
  grub2-mkconfig (8)  - Generate a GRUB configuration file.

# grub2-mkconfig -o /boot/grub2/grub.cfg
```

```
Generating grub configuration file ...
done

# reboot
```

再次检查一下,示例命令如下:

```
# virt-host-validate
  QEMU: Checking for hardware virtualization                              : PASS
  QEMU: Checking if device /dev/kvm exists                                : PASS
  QEMU: Checking if device /dev/kvm is accessible                         : PASS
  QEMU: Checking if device /dev/vhost-net exists                          : PASS
  QEMU: Checking if device /dev/net/tun exists                            : PASS
  QEMU: Checking for cgroup 'cpu' controller support                      : PASS
  QEMU: Checking for cgroup 'cpuacct' controller support                  : PASS
  QEMU: Checking for cgroup 'cpuset' controller support                   : PASS
  QEMU: Checking for cgroup 'memory' controller support                   : PASS
  QEMU: Checking for cgroup 'devices' controller support                  : PASS
  QEMU: Checking for cgroup 'blkio' controller support                    : PASS
  QEMU: Checking for device assignment IOMMU support                      : PASS
  QEMU: Checking if IOMMU is enabled by Kernel                            : PASS
```

至此,我们的实验环境就准备好了。

2.3 KVM 的管理方法

根据管理员所处位置的不同,KVM 的管理分为两种形式:

(1)本地管理。
(2)远程管理。

2.3.1 本地管理

所谓本地管理就是指管理员坐在计算机面前直接操作,最常见的场景是桌面环境或还未配置网络的服务器环境。

对于我们实验环境来讲,如果直接在 VMware Workstation Pro 或 Player 软件中进行操作,则相当于本地管理。

KVM 本地管理既可以通过命令行命令实现,也可以通过 GUI 下的管理工具实现。

virsh 是一个功能强大的命令行虚拟化管理工具,virt-install 是一个安装新虚拟机的命令行工具,如图 2-31 所示。

Linux 图形环境准确就绪之后(例如在命令提示符中执行了 startx 命令),就可以在应用程序中找到 Virtual Machine Manager 的图标,如图 2-32 所示。双击这个图标,就可以启动它了,如图 2-33 所示。

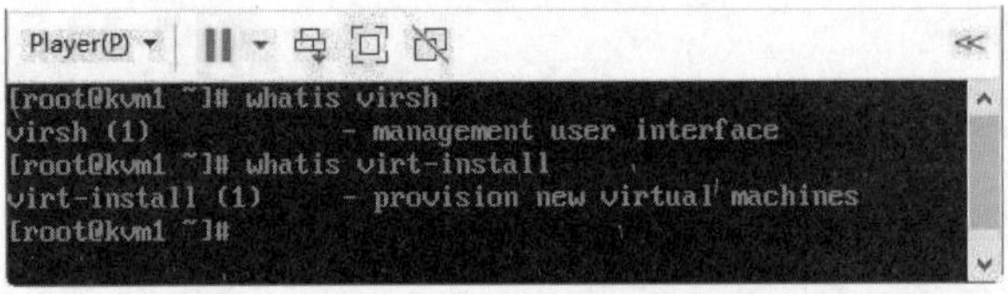

图 2-31 命令行管理

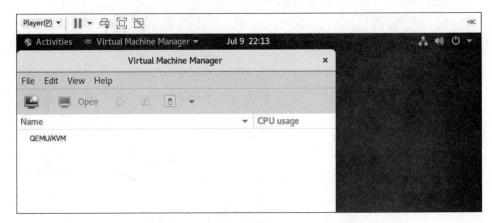

图 2-32 图形界面中的应用程序

图 2-33 图形化的虚拟机管理工具 Virtual Machine Manager

2.3.2 远程管理

远程管理是一种在不直接接触计算机系统的条件下的管理方式。用户使用特定的软件通过网络登录计算机系统,然后进行管理和控制。

对 Linux 进行远程管理的方法有很多,下面介绍 5 种在 Windows 操作系统中远程管理 Linux 主机的方法。

1. 通过 SSH 进行远程管理

Linux 最常见的远程管理方式是使用 SSH(Secure Shell)协议,它是一种作用于应用层上的 C/S 架构的安全协议。常见的 SSH 服务器端组件是 OpenSSH。Windows 下的 SSH 客户端软件有很多,在后续的实验中我们将使用开源的 PuTTY 软件。

在 PuTTY 界面选择 Session,如图 2-34 所示,在 Host Name (or IP address) 文本框中输入远程主机的 IP 地址。端口号(Port)根据使用的协议有所区别,SSH 默认使用 22。在 Protocol 中选择使用的协议 SSH,然后单击 Open 按钮,登录之后就可以连接 Linux 主机了。

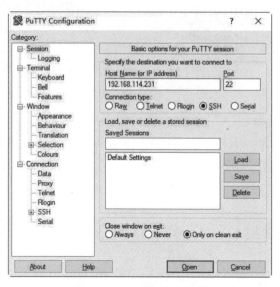

图 2-34　PuTTY 的主界面

通过 SSH 进行 KVM 远程管理与本地管理类似,如图 2-35 所示。

2. 通过 VNC 进行远程管理

VNC 是虚拟网络控制台(Virtual Network Console)的缩写,是一款优秀的远程控制工具软件。它远程控制能力强大、高效实用,而且是跨平台的。

VNC 分为服务器端和客户端两部分,服务器端组件 VNC Server 可以运行在 Linux、UNIX、Windows 等多种操作系统上,客户端组件 VNC Viewer 也有多种平台上的版本,如图 2-36 所示。

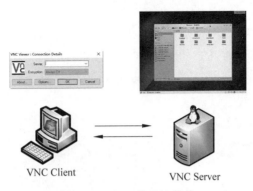

图 2-35　执行 KVM 的管理命令

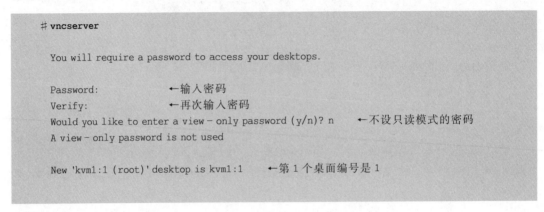

图 2-36　VNC 的 C/S 架构

1）服务器端软件安装

在 CentOS 8 中包含一个名为 TigerVNC Server 的服务器端软件，首先我们在 Linux 主机上安装服务器端组件，命令如下：

```
# dnf install tigervnc-server
```

如果是早期的 TigerVNC Server 版本，可以执行 vncserver 命令来启动服务器端进程，命令如下：

```
# vncserver

  You will require a password to access your desktops.

  Password:              ←输入密码
  Verify:                ←再次输入密码
  Would you like to enter a view-only password (y/n)? n   ←不设只读模式的密码
  A view-only password is not used

  New 'kvm1:1 (root)' desktop is kvm1:1    ←第 1 个桌面编号是 1
```

```
Creating default startup script /root/.vnc/xstartup
Creating default config /root/.vnc/config
Starting applications specified in /root/.vnc/xstartup
Log file is /root/.vnc/kvm1:1.log
```

如果是新版本的 TigerVNC Server,则在执行 vncserver 命令时会出现如下提示信息:

```
# vncserver
    vncserver has been replaced by a systemd unit.
    Please read /usr/share/doc/tigervnc/HOWTO.md for more information.
```

根据/usr/share/doc/tigervnc/HOWTO.md 这个 Markdown 格式文件的说明,可以采用如下操作进行配置。

(1) 添加用户映射。

```
# vi /etc/tigervnc/vncserver.users
    # 在最后添加
    :1=root
```

VNC 服务器会监听 5900+桌面编号,例如配置 root 用户使用 1 号桌面,那么 VNC 服务器会监听 5901 端口。

(2) 配置 Xvnc 选项。

```
# vi /etc/tigervnc/vncserver-config-defaults
    在最后添加
    session=gnome
```

(3) 为 root 用户设置 VNC 密码。

```
# id -un
    root

# vncpasswd
    Password: 输入密码
    Verify: 再次输入密码
    Would you like to enter a view-only password (y/n)? n
    A view-only password is not used
```

(4) 启动 Tigervnc 服务器。

```
# systemctl start vncserver@:1.service

# systemctl enable vncserver@:1.service
```

我们还需要配置防火墙以便允许 VNC 客户端进行访问。VNC 服务器的端口是动态端口,它会在 5900＋桌面编号这个范围内变化,所以最简洁的方法是在防火墙策略中添加 VNC 服务,命令如下:

```
#firewall-cmd --info-service=vnc-server
  vnc-server
    ports: 5900-5903/tcp
    protocols:
    source-ports:
    modules:
    destination:
    includes:

#firewall-cmd --add-service=vnc-server --permanent

#firewall-cmd --reload

#firewall-cmd --list-services
  cockpit dhcpv6-client ssh vnc-server
```

2)客户端软件安装与使用

Windows 下常见的 VNC 客户端有 2 种:Realvnc viewer 和 Tightvnc viewer。下面我们使用 Realvnc viewer 来做实验。

首先,从 https://www.realvnc.com/download/vnc/下载相应版本的客户端软件,然后进行安装。

安装过程很简单。安装之后启动客户端,在网址栏中输入远程桌面的标识。远程桌面的标识有两种格式。

第一种格式:IP＋冒号＋桌面编号。如:192.168.114.231:1。

第二种格式:IP＋冒号＋TCP 端口号编号。如:192.168.114.231:5901。

按第一种格式输入 192.168.114.231:1,然后按回车键,如图 2-37 所示。VNC Viewer 会进行安全检查,如果服务器与客户机之间通信是不加密的,就会发出警告,如图 2-38 所示,单击 continue 按钮继续。

输入前面所设置的密码之后,如图 2-39 所示,我们可以访问远程主机的图形界面了,如图 2-40 所示。

3.通过 XRDP 进行远程管理

还有一种与 VNC 类似的远程桌面访问方法是 XRDP。XRDP 是一个开源的远程桌面协议(Remote Desktop Protocol,RDP)服务器,采用的是标准的 RDP。它的网址是 http://xrdp.org/。

VNC 默认需要设置独立的密码,而 XRDP 服务器则直接使用 Linux 操作系统的账户,所以管理会比较方便一些。

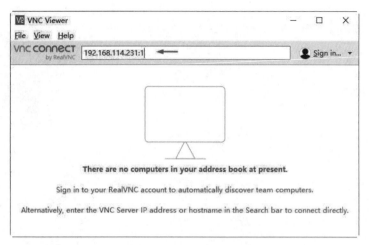

图 2-37　VNC Viewer 的主界面

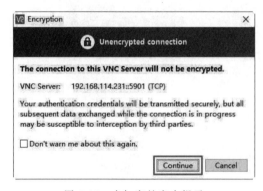

图 2-38　未加密的安全提示

图 2-39　输入所设置的密码

XRDP 的客户端软件，除了我们常用的微软的远程桌面客户端之外，还可以使用 FreeRDP、rdesktop、NeutrinoRDP 等客户端软件。

1）服务器端软件安装

CentOS 8 自带的 3 个软件仓库 AppStream、BaseOS 和 Extras 中并没有 XRDP，它在 EPEL 软件仓库中，所以需要先安装 EPEL 软件仓库的配置，命令如下：

```
# dnf repolist
  repo id          repo name
  AppStream        CentOS-8 - AppStream
  BaseOS           CentOS-8 - Base
  extras           CentOS-8 - Extras

# dnf -y install epel-release
```

图 2-40 远程主机的图形界面

```
# dnf repolist
  repo id          repo name
  AppStream        CentOS-8 - AppStream
  BaseOS           CentOS-8 - Base
  epel             Extra Packages for Enterprise Linux 8 - x86_64
  epel-modular     Extra Packages for Enterprise Linux Modular 8 - x86_64
  extras           CentOS-8 - Extras
```

提示：EPEL 是企业版 Linux 的额外软件包（Extra Packages for Enterprise Linux）的缩写，它是 Fedora 小组维护的一个软件仓库项目。

接下来就可以安装 XRDP 了，并且可以设置为自动启动，命令如下：

```
# dnf -y install xrdp

# systemctl start xrdp

# systemctl enable xrdp
```

最后还需要配置防火墙，允许 RDP（TCP 3389 端口）入站，命令如下：

```
#firewall-cmd --add-service=rdp --permanent

#firewall-cmd --reload
```

2）客户端连接

我们可以使用 Windows 中的远程桌面客户端进行连接，如图 2-41 所示。输入 Linux 操作系统中的用户名和密码就可以访问远程主机的图形界面了，如图 2-42 所示。

图 2-41　远程桌面客户端软件

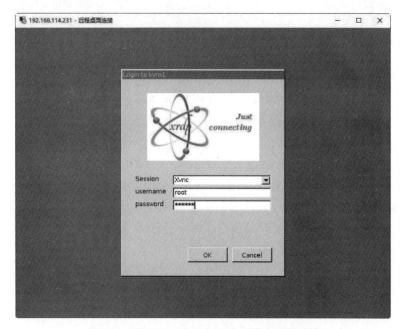

图 2-42　输入 Linux 操作系统的用户名与密码

以这种方式访问 Linux 主机与 VNC 访问 Linux 主机很类似,在 Application 中双击 Virtual Machine Manager 图标即可启动虚拟机管理器,如图 2-43 所示。

图 2-43　远程主机的图形界面

提示:普通的 Linux 用户也可以登录 XRDP。

---

4. 通过 X-Window 进行远程管理

如果仅仅想单独使用 virt-manager、virt-viewer 等图形管理工具而不访问整个桌面,则常用的做法是通过 SSH 和 X-Window 组合实现。

X Window System 简称为 X、X11 或 X-Window。它是一个开源、跨平台的客户端-服务器计算机软件系统,可在分布式网络环境中提供 GUI。它的功能很强大,包括网络透明性、可定制的图形功能等特性。Linux 发行版本中使用得最多的 Gnome 和 KDE 桌面都是基于 X-Window 系统而构建出来的。

X-Window 采用的是客户机/服务器架构,服务器接收客户机的请求来绘制窗口、图形,同时它将来自鼠标、键盘等输入设备的信息传递给客户机。

X-Window 的客户机与服务器组件,既可以在同一台主机上,也可以分布在不同的主机上,如图 2-44 所示。左边这台微软 Windows 计算机安装有 X-Window 的 Server 端软件,那么它就是服务器。右边这台 Linux 的主机如果将绘制窗口或图形的请求发送给左边这台机器,那么这就是客户机。

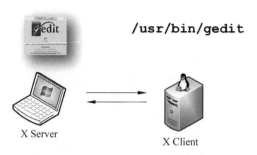

图 2-44　X-Window 的 C/S 架构

这时,如果在 Linux 的主机上运行/usr/bin/gedit 命令,由于这是图形化的应用程序,所以它会把软件窗口等图形信息发送给左边这台 Windows 主机上的 X Server 组件。同时,如果在 Windows 主机上移动鼠标、敲击键盘,这些输入信息都会转发到 Linux 机器上。

X-Window 的客户机与服务器组件如果位于不同主机上,就要考虑两者通信的安全性了。最常用的方法是先在两台主机之间通过 SSH 构建一条安全通道,它就像一条安全的隧道,黑客无法获得其中传递的真实数据,然后,在其中传递 X-Window 的流量,如图 2-45 所示。

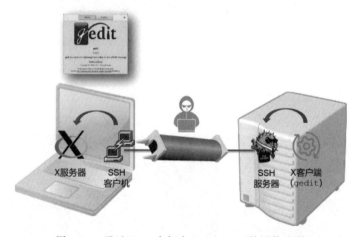

图 2-45　通过 SSH 来加密 X-Window 的网络流量

具体的实现过程:首先需要在左边 Windows 主机上安装 X-Window 服务器,例如 Xming。Xming 是一个运行在 Windows 系统中的开源 X Window Server,它的下载网址是 http://sourceforge.net/projects/xming/。下载并安装之后启动 Xming,这台计算机就成为一台 X-Window 服务器,这时会在托盘中出现一个小图标,如图 2-46 所示。

图 2-46　Xming

然后还需要在 SSH 客户端上进行配置，例如在 PuTTY 上进行设置，如图 2-47 所示。

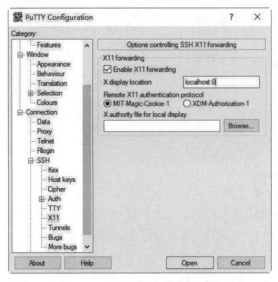

图 2-47　在 PuTTY 中配置 X11 转发

依次单击 Connection、SSH、X11，选中 Enable X11 forwarding，然后将 X display location 设置为 localhost:0。这里的 0 就是在 Xming 中所设置的号码，再单击 Open 按钮连接到 Linux 主机。

在 Linux 主机上执行一个需要 X-Window 的 GUI 程序，例如 virt-manager，如图 2-48 所示。

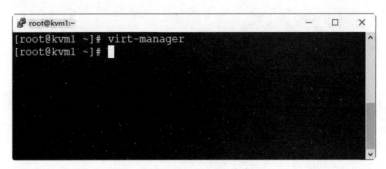

图 2-48　执行 GUI 的应用程序

这个 virt-manager 就是一个 X 客户端，它会将窗口等图形数据转发给 SSH 服务器，SSH 服务器又将其转发给 SSH 客户端，即 PuTTY，PuTTY 再转发给 X 服务器。这样，virt-manager 就显示在我们这台机器的屏幕上了，如图 2-49 所示。

5. 通过 Cockpit 进行远程管理

Cockpit 是红帽公司力推的一个开源的基于 Web 的 Linux 管理工具。它的网址是

图 2-49 X 服务器上显示的 GUI 应用程序

https://cockpit-project.org。

Cockpit 字面的意思是驾驶员座舱，它提供一个友好的 Web 前端界面，非常简单易用。Linux 系统管理员通过它可以执行诸如存储管理、网络配置、检查日志、管理虚拟机、容器等任务。另外，还可以通过 Cockpit 实现多台服务器集中式管理。

1）服务器端软件安装

在安装 CentOS 8 时，如果选中了 Remote Management for Linux 这个软件包组，则会自动安装 Cockpit。当然也可以手工进行安装并设置为自动启动，命令如下：

```
# dnf -y install cockpit

# systemctl start cockpit.socket

# systemctl enable cockpit.socket
```

安装 Cockpit 软件包的时候，默认会自动修改防火墙，添加 Cockpit 使用的 TCP 9090 端口。当然，我们也可以通过 firewall-cmd 命令自行添加，命令如下：

```
# firewall-cmd --info-service=cockpit
 cockpit
   ports: 9090/tcp
   protocols:
   source-ports:
   modules:
   destination:
   includes:
   helpers:

# firewall-cmd --add-service=cockpit --permanent

# firewall-cmd --reload
```

Cockpit 提供了插件接口以方便功能的扩展。可以通过 dnf 命令在软件仓库中搜索与

Cockpit 有关的软件包，命令如下：

```
# dnf search cockpit
========== Name Exactly Matched: cockpit ========================
cockpit.x86_64 : Web Console for Linux servers
========== Name & Summary Matched: cockpit ======================
cockpit-ws.x86_64 : Cockpit Web Service
cockpit-pcp.x86_64 : Cockpit PCP integration
cockpit-composer.noarch : Composer GUI for use with Cockpit
cockpit-doc.noarch : Cockpit deployment and developer guide
cockpit-bridge.x86_64 : Cockpit bridge server-side component
cockpit-session-recording.noarch : Cockpit Session Recording
cockpit-packagekit.noarch : Cockpit user interface for packages
cockpit-dashboard.noarch : Cockpit remote servers and dashboard
cockpit-podman.noarch : Cockpit component for Podman containers
cockpit-machines.noarch : Cockpit user interface for virtual machines
subscription-manager-cockpit.noarch : Subscription Manager Cockpit UI
cockpit-storaged.noarch : Cockpit user interface for storage, using udisks
cockpit-system.noarch : Cockpit admin interface package for configuring and troubleshooting a system
```

为了后续实验方便，建议添加存储管理和虚拟化管理的插件，命令如下：

```
# dnf -y install cockpit-machines cockpit-storaged
```

2）客户端（浏览器）访问

下面，我们就可以通过对 HTML5 支持比较好的浏览器（例如 Chrome 或 Firefox 浏览器）访问 Cockpit 了。访问的 URL 格式为 https://服务器的 IP 地址:9090/。

由于 Cockpit 默认使用的是自签名的证书，所以有些浏览器会有安全警告，如图 2-50(a)所示。单击"高级"按钮，然后单击"继续前往 192.168.114.231(不安全)"继续访问此网站，如图 2-50(b)所示。

在登录页面中输入 Linux 系统中的用户名及密码，这样就可以访问 Cockpit 了，如图 2-51 所示。

提示：普通的 Linux 用户也可以登录 Cockpit。

Cockpit 有一个很友好的交互式服务器管理界面，无须阅读帮助文件就可以发现如何通过 Cockpit 执行任务，如图 2-52 所示。单击"虚拟机"，就可以进入虚拟机管理的页面，如图 2-53 所示。

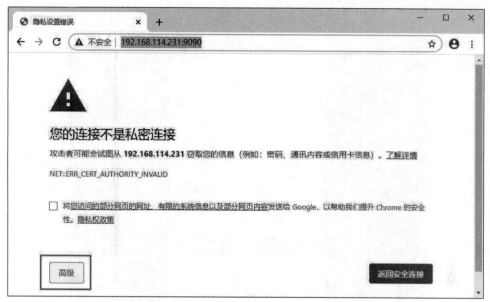

(a)

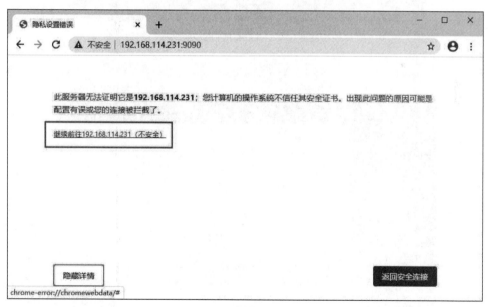

(b)

图 2-50 自签名网站的安全警告

图 2-51　Cockpit 的登录页面

图 2-52　Cockpit 的概览页面

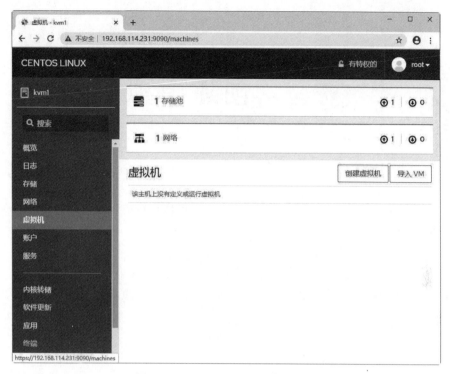

图 2-53　Cockpit 的虚拟机管理页面

2.4　本章小结

本章讲解了如何安装配置 KVM 虚拟化主机，掌握了 KVM 虚拟化都需要安装哪些软件包，了解了验证虚拟化功能状态的方法及管理 KVM 的工具。

第 3 章将通过示例讲解如何通过 Cockpit、virt-manager 和 virt-install 创建虚拟机，以及在虚拟机中安装半虚拟化驱动 VirtIO 及 Agent 的方法。

第 3 章 创建虚拟机

第 2 章讲解了 KVM 相关组件的安装，本章将讲解虚拟机的创建。在 RHEL/CentOS 8 中创建虚拟机的方法主要有 3 种：Cockpit、virt-manager 和 virt-install。其中前两种方法是基于 GUI 的，virt-install 则是在命令行进行安装。

新用户总喜欢 GUI 而不喜欢命令行，因此我们会先从 Cockpit、virt-manager 开始来创建虚拟机，但同时也会尽可能地通过 virsh 来观察背后所发生的变化，这样才能深入地掌握所学习的知识。

本章要点：
- 使用 Cockpit 创建虚拟机。
- 使用 virt-manager 创建虚拟机。
- 使用 virt-install 创建虚拟机。
- 半虚拟化驱动 VirtIO。
- QEMU Guest Agent。
- 显示设备与协议原理。
- SPICE Agent。

3.1 使用 Cockpit 创建虚拟机

按 Fedora 社区和红帽公司的规划，将来会逐步用 Cockpit 替代 virt-manager 来管理虚拟机。虽然目前 Cockpit 的功能还比不上 virt-manager 丰富，但是基于 Web 的管理模式是大势所趋。

3.1.1 查看当前配置

在创建虚拟机之前，我们先通过 Cockpit 和 virsh 来查看当前的配置。

Cockpit 虚拟化管理的主界面中会显示虚拟机列表、存储池的数量及虚拟网络的数量，如图 3-1 所示。目前系统没有虚拟机，但是有一个默认的存储池和虚拟网络。

图 3-1 Cockpit 虚拟化管理的主界面

什么是存储池？libvirt 通过存储池（Storage Pool）和卷（Volume）在宿主机上提供存储管理。存储池是管理员（通常是专门的存储管理员）留出的一定空间的存储，供虚拟机使用。存储池由存储管理员或系统管理员划分为多个存储卷，并将这些卷作为块设备分配给虚拟机。在本书的第 6 章会详细讲解这些知识。

单击"存储池"链接，就会出现一个存储池清单，如图 3-2 所示。这个名为 default 的存储池的信息如下：

图 3-2 Cockpit 中的存储池

(1) 存储名称：default。
(2) 状态：已激活。
(3) 已分配：3.85GB。
(4) 容量：49.98GB。
(5) 目标路径：/var/lib/libvirt/images。
(6) 持久：是。
(7) 自动启动：是。
(8) 类型：dir。

virsh 是管理 KVM 虚拟化最主要的管理工具。它功能强大，除 KVM 之外，还可以管理 Xen、LXC、OpenVZ、VirtualBox 和 VMware ESXi/ESX 等多种虚拟化平台。

virsh 有很多种子命令，其中与存储池有关的子命令通常以 pool-开头。先简单了解一下 3 个子命令。

(1) pool-list：列出当前的存储池。默认仅显示处于激活状态的存储池。
(2) pool-info：返回特定存储池对象的基本信息。
(3) pool-dumpxml：返回特定存储池对象的详细信息，输出格式为 XML。

查看存储池 default 的命令如下：

```
# virsh pool-list
Name                 State      Autostart
-------------------------------------------
default              active     yes

# virsh pool-info default
Name:           default
UUID:           b111d4c0-374c-4d8f-bf58-1050e4af953d
State:          running
Persistent:     yes
Autostart:      yes
Capacity:       49.98 GiB
Allocation:     3.85 GiB
Available:      46.13 GiB

# virsh pool-dumpxml default
<pool type='dir'>
  <name>default</name>
  <uuid>b111d4c0-374c-4d8f-bf58-1050e4af953d</uuid>
  <capacity unit='Bytes'>53660876800</capacity>
  <allocation unit='Bytes'>4129992704</allocation>
  <available unit='Bytes'>49530884096</available>
  <source>
  </source>
  <target>
```

```
    <path>/var/lib/libvirt/images</path>
    <permissions>
        <mode>0711</mode>
        <owner>0</owner>
        <group>0</group>
        <label>system_u:object_r:virt_image_t:s0</label>
    </permissions>
  </target>
</pool>
```

类似地,我们再使用 Cockpit 和 virsh 对照着查看虚拟网络的信息。现在,在 Cockpit 的虚拟网络中仅显示了一个名为 default 的虚拟网络,如图 3-3 所示。它的主要信息如下:

(1) 名称:default。

(2) 设备:virbr0。

(3) 转发模式:NAT。

(4) 状态:激活。

(5) IP 地址:192.168.122.1。

(6) DHCP 范围:192.168.122.2-192.168.122.254。

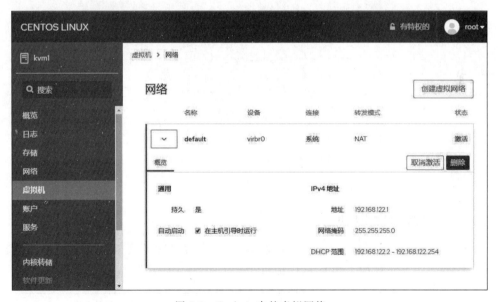

图 3-3　Cockpit 中的虚拟网络

virsh 中与宿主机网络有关的子命令通常以 net-开头。先简单了解一下 3 个子命令。

(1) net-list:列出当前的网络,默认仅显示处于激活状态的网络。

(2) net-info:返回特定网络对象的基本信息。

(3) net-dumpxml:返回特定网络对象的详细信息,输出格式为 XML。

查看虚拟网络 default 的命令如下：

```
# virsh net-list
Name          State       Autostart    Persistent
----------------------------------------------------
default       active      yes          yes

# virsh net-info default
Name:           default
UUID:           52959885-1cb7-425e-ae2d-4de2f98f02bc
Active:         yes
Persistent:     yes
Autostart:      yes
Bridge:         virbr0

# virsh net-dumpxml default
<network>
  <name>default</name>
  <uuid>52959885-1cb7-425e-ae2d-4de2f98f02bc</uuid>
  <forward mode='nat'>
    <nat>
      <port start='1024' end='65535'/>
    </nat>
  </forward>
  <bridge name='virbr0' stp='on' delay='0'/>
  <mac address='52:54:00:fd:b2:60'/>
  <ip address='192.168.122.1' netmask='255.255.255.0'>
    <dhcp>
      <range start='192.168.122.2' end='192.168.122.254'/>
    </dhcp>
  </ip>
</network>
```

3.1.2 创建虚拟机

如果在嵌套虚拟化环境做实验，则建议使用 32 位版的操作系统以减少资源开销。下面将创建、安装一个 32 位版 CentOS 的虚拟机。

提示：由于从版本 7 开始，CentOS 就不再提供 32 位版的 ISO 文件了，所以我们下载 CentOS 6 的 32 位版本，例如 CentOS 6.10 的最小化版本。

从 CentOS 官方网站下载 CentOS-6.10-i386-minimal.iso，并保存在 /iso 目录中，命令如下：

```
# mkdir /iso

# cd /iso

# wget \
http://mirrors.163.com/centos/6.10/isos/i386/CentOS-6.10-i386-minimal.iso

# ls -l
total 364544
-rw-r--r--. 1 root root 373293056 Jun 30 2018 CentOS-6.10-i386-minimal.iso

# file CentOS-6.10-i386-minimal.iso
CentOS-6.10-i386-minimal.iso: DOS/MBR boot sector; partition 1 : ID=0x17, active, start
-CHS (0x0,0,1), end-CHS (0x163,63,32), startsector 0, 729088 sectors
```

在虚拟机页面中单击"创建虚拟机"按钮，就会出现一个"创建新的虚拟机"窗口，如图 3-4 所示。

图 3-4 在 Cockpit 中创建新的虚拟机

创建新的虚拟机，需要提供这样一些选项：
1. 名称
虚拟机的名称，例如 centos6.10。

2．安装类型

有这样几种选项：

（1）下载 OS：从 Cockpit 的操作系统仓库中下载操作系统。

（2）本地安装介质：使用已经下载好的安装介质。

（3）URL：从操作系统安装介质树的 URL 进行安装。

（4）网络引导（PXE）：通过网络引导进行安装。

本次实验将选中"本地安装介质"，然后在安装源中指定/iso/目录中已经下载好的 ISO 文件。

3．操作系统

指定要安装的操作系统类型。从下拉列表中选择 CentOS 6.10。

4．存储来源及大小

从下拉列表中选择"创建新卷"选项。指定磁盘大小，保持默认的 9GiB 即可。

5．内存

指定内存大小，保持默认的 1GB 即可。

选中"立即启动 VM"，然后单击"创建"按钮。这样便会创建一个新的虚拟机并启动，然后会自动切换到"控制台"界面，如图 3-5 所示。

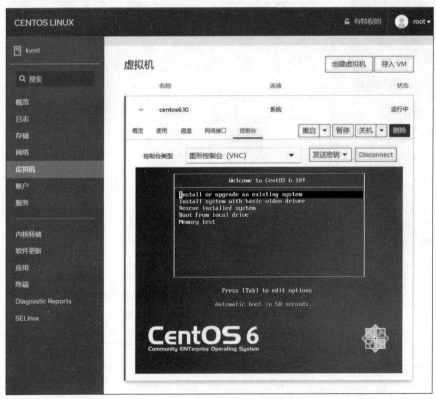

图 3-5　Cockpit 中虚拟机的控制台

由于此时默认的引导次序是安装介质优先,所以会启动 CentOS 6.10 的安装程序。这时,要确保将"图形控制台(VNC)"作为控制台类型。接下来,就可以根据具体的提示来安装操作系统了。

安装结束后,单击 Reboot 按钮重新启动虚拟机,如图 3-6 所示。

图 3-6　虚拟机安装完毕

3.1.3　查看虚拟机与环境的配置

下面,我们再通过 Cockpit 和 virsh 命令查看虚拟机与环境的配置。

在 Cockpit 的虚拟机主界面中会看到新创建的虚拟机,如图 3-7 所示。

图 3-7　虚拟机列表

virsh 的 list 子命令也可以获得虚拟机列表，登录命令如下：

```
# virsh list --all
 Id    Name                           State
----------------------------------------------------
 1     centos6.10                     running
```

执行命令后会列出全部虚拟机。如果不使用--all 选项，则仅列出已启动的虚拟机。在 libvirt 中，虚拟机对应的术语是域（domain）。

在输出中，ID 是虚拟机标识，Name 是虚拟机的名称。在 virsh 命令中，既可以通过 ID 也可以通过 Name 来引用虚拟机。State 是虚拟机的运行状态，共分为 7 种状态：

（1）running：运行中。
（2）idle：空闲。
（3）paused：已暂停。
（4）in shutdown：正在关机。
（5）shut off：已关闭。
（6）crashed 崩溃。
（7）pmsuspended 暂停。

在 Cockpit 中，单击虚拟机名称左边的"ˇ"链接，会显示此虚拟机的概要信息，如图 3-8 所示。

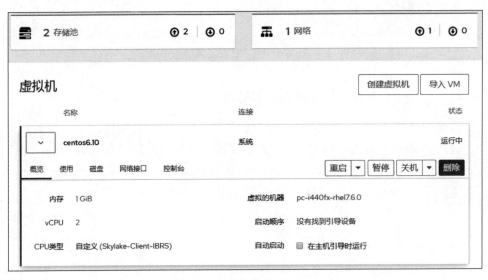

图 3-8 虚拟机的详细信息

对应的 virsh 子命令是 dominfo。在使用时，需要为 dominfo 子命令提供虚拟机的 ID 或名称，命令如下：

```
# virsh dominfo 1
Id:              1
Name:            centos6.10
UUID:            edcba2b1-afa9-4de5-8de4-b42cc1bc66c9
OS Type:         hvm
State:           running
CPU(s):          2
CPU time:        40.7s
Max memory:      1048576 KiB
Used memory:     1048576 KiB
Persistent:      yes
Autostart:       disable
Managed save:    no
Security model:  seLinux
Security DOI:    0
Security label:  system_u:system_r:svirt_t:s0:c168,c621 (enforcing)
```

提示：virsh 的子命令名称及格式很有规律，例如：dom 是 domain 的缩写，info 是 information 的缩写，掌握这个规律可以大大提高效率。

与 Cockpit 中"使用"选项卡对应的子命令是 domstats，stats 是 statistics 的缩写，示例命令如下：

```
# virsh domstats 1
```

与 Cockpit 中"磁盘"选项卡对应的子命令是 domblklist，blk 是 blocks 的缩写，示例命令如下：

```
# virsh domblklist 1
```

与 Cockpit 中"网络接口"选项卡对应的子命令是 domiflist，if 是 interfaces 的缩写，示例命令如下：

```
# virsh domiflist 1
```

提示：可以通过 virsh help 命令来获得联机帮助，例如 virsh help domiflist。另外，virsh 命令有类似于 Bash Shell 的自动补全功能，使用 Tab 键可以对子命令、选项和参数进行自动补全。

查看完虚拟机，下面查看宿主机的存储。

在 Cockpit 中，单击 default 存储池中的"存储卷"，会发现有一个新的存储卷，这就是新创建的虚拟机的虚拟磁盘,如图 3-9 所示。

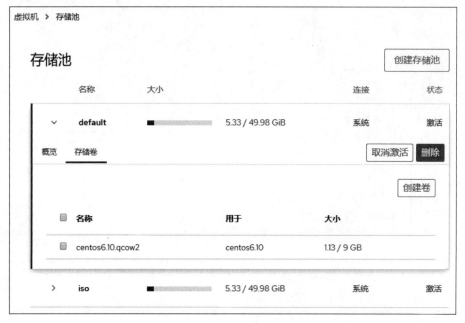

图 3-9　存储池 default 中的存储卷

对应的 virsh 子命令是 vol-list,在使用时需要提供存储池的标识,命令如下:

```
# virsh vol-list default
Name                 Path
-----------------------------------------------------------
centos6.10.qcow2     /var/lib/libvirt/images/centos6.10.qcow2
```

在 Cockpit 中,还会发现 1 个名为 iso 的新存储池,如图 3-10 所示。

在这个存储池中有 1 个存储卷,就是在安装 centos6.10 虚拟机时使用的 ISO 文件,如图 3-11 所示。

virsh 子命令 pool-list 和 pool-info 可以获得存储池的列表和特定存储池的信息,命令如下:

```
# virsh pool-list
Name        State      Autostart
-------------------------------------------
default     active     yes
iso         active     yes


# virsh pool-info iso
Name:       iso
UUID:       51c3aa8a-6df2-4a46-b121-81bd44f04298
```

图 3-10 新增加的存储池

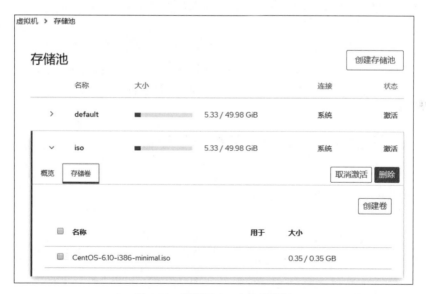

图 3-11 存储池 iso 中的存储池

```
State:          running
Persistent:     yes
Autostart:      yes
Capacity:       49.98 GiB
Allocation:     5.33 GiB
Available:      44.65 GiB
```

virsh 子命令 vol-list 可以查看指定存储池中的存储卷列表。在使用时需要指定存储池的名称，命令如下：

```
# virsh vol-list iso
Name                              Path
----------------------------------------------------------
CentOS-6.10-i386-minimal.iso      /iso/CentOS-6.10-i386-minimal.iso
```

最后，我们使用 nmcli 观察宿主机上网络连接的变化，命令如下：

```
# nmcli connection show
NAME     UUID                                  TYPE      DEVICE
ens32    0b1638a6-add5-4057-9bce-575efc3d5bf2  ethernet  ens32
virbr0   fc81f0db-11eb-4471-84b6-3bb6b0c44f7b  bridge    virbr0
vnet0    27ebc3da-dc3d-4a9c-8880-6ce52af0cf8f  tun       vnet0
```

在输出中，我们会看到新增了一个名为 vnet0 的网络设备，它与虚拟机的网卡相连接。

3.2 使用 virt-manager 创建虚拟机

虚拟机管理器 virt-manager 是一个管理宿主机和虚拟机的 GUI 管理工具。虽然 RHEL/CentOS 8 还包含这个软件，但是会在将来用 Cockpit 替换它。可是目前比较"尴尬"的地方是 Cockpit 功能并不完整，有些功能只能通过 virt-manager 或 virsh 来完成，所以我们还必须掌握 virt-manager 的使用。

3.2.1 使用 virt-manager 查看当前配置

第 2 章介绍了在 VNC、XRDP 和 X-Window 等 3 种环境中启动 virt-manager 的方法，不管采用哪种方法启动，都会先看到一个虚拟机列表，如图 3-12 所示。

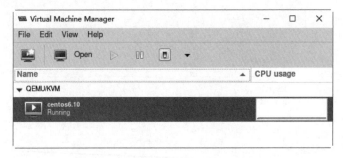

图 3-12　virt-manager 的主界面

双击虚拟机的名称或者单击工具栏中的 ▶ 图标，就会显示此虚拟机控制台和细节信息，可以在控制台中管理这台虚拟机，如图 3-13 所示。

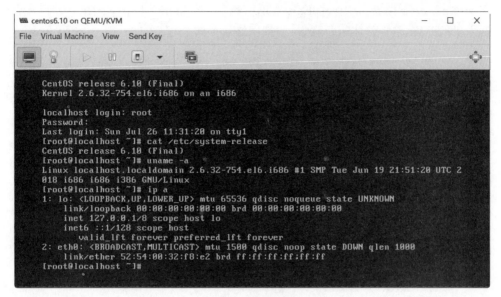

图 3-13 virt-manager 中的虚拟机控制台

单击工具栏中的 图标,会显示此虚拟机的详细配置,可以在这里查看和修改虚拟机的配置,如图 3-14 所示。

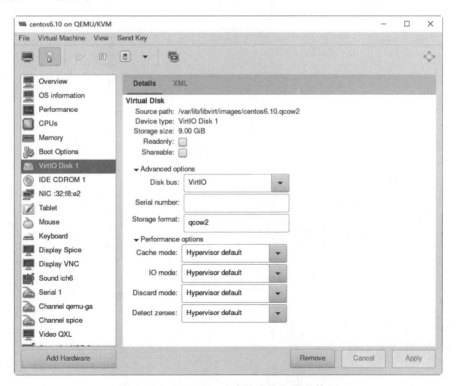

图 3-14 virt-manager 中的虚拟机配置管理

在 virt-manager 主界面中单击 Edit 菜单中的 Connection Details，如图 3-15 所示，就会显示宿主机的配置，包括整体运行状态、虚拟网络和存储，如图 3-16～图 3-18 所示。

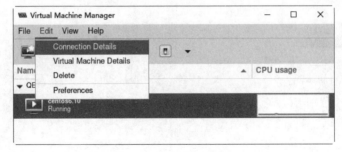

图 3-15　在 virt-manager 中打开连接的细节

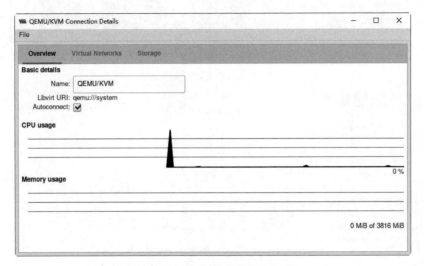

图 3-16　宿主机的整体运行状态

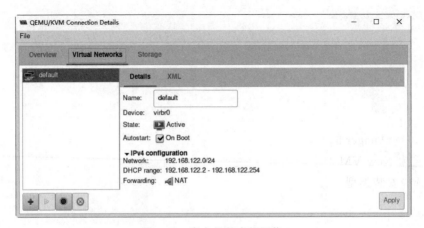

图 3-17　宿主机的虚拟网络

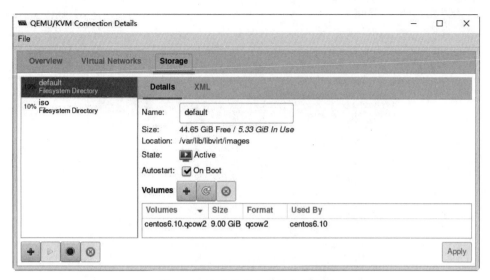

图 3-18　宿主机的存储池与存储卷

3.2.2　创建虚拟机

与前面的实验类似，为了减少资源开销我们将安装一个 32 位版的 Windows Server 2003 虚拟机。首先将 ISO 文件通过 WinSCP 类软件上传到宿主机/iso 目录中，如图 3-19 所示。

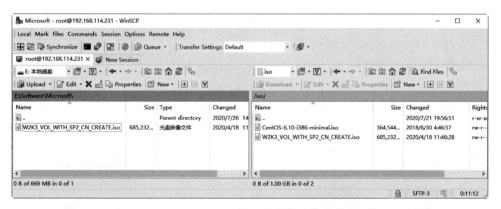

图 3-19　将 Windows Server 2003 的 ISO 文件上传到宿主机的/iso 目录中

单击 virt-manager 的 File 菜单中的 New Virtual Machine 或工具栏中的 图标，就会出现一个名为 New VM 的向导。它将虚拟机的创建过程分为 5 步：
(1) 选择安装类型。
(2) 查找和配置安装介质。
(3) 配置内存和 CPU 选项。

（4）配置虚拟机的存储。

（5）配置虚拟机名称、网络、体系结构和其他硬件设置。

第 1 步需要在 4 种安装操作系统的方式中选择一种，如图 3-20 所示。

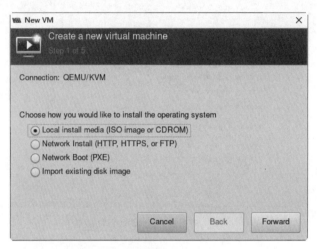

图 3-20　选择安装方式

1）Local install media（ISO image or CDROM）

此方法使用 ISO 格式的映像文件。虽然写有 CDROM，但是目前还是无法通过宿主机上的 CD-ROM 或 DVD-ROM 设备进行安装。

2）Network Install（HTTP，HTTPS，or FTP）

通过保存在 HTTP、HTTPS 或 FTP 服务器上的操作系统安装文件来安装。如果选择此项，还需要提供安装文件的 URL 及内核选项。

3）Network Boot（PXE）

采用预引导执行环境（PXE）服务器来安装虚拟机。

4）Import existing disk image

创建新的虚拟机，并将现有的磁盘映像（包含预安装的可引导操作系统）导入该虚拟机。

现在我们选中 Local install media（ISO image or CDROM）并单击 Forward 按钮继续。

第 2 步需要查找和配置安装介质，如图 3-21 所示。单击 Browse 按钮打开 Choose Storage Volume 窗口，如图 3-22 所示。

由于之前通过 Cockpit 创建虚拟机时已经创建过一个名为 iso 的存储池，所以可以在其中找到 Windows Server 20003 的 ISO 文件，然后单击 Choose Volume 按钮，将返回 New VM 向导。

接下来，需要设置操作系统类型和版本信息。为虚拟机设置适当的操作系统类型及版本，是其流畅工作的前提。如果选中了 Automatically detect from the installation media/source 复选框，则向导会自动检测安装介质并设置操作系统类型与版本，如图 3-23 所示。如果没有检测出或者检测的结果不正确，我们还可以手动指定。单击 Forward 按钮继续。

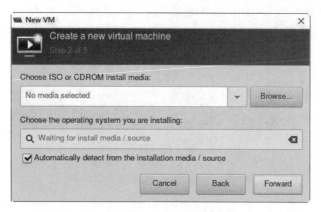

图 3-21　从本地 ISO 映像安装

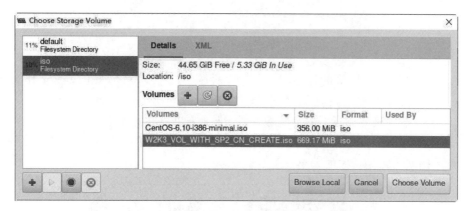

图 3-22　选择存储卷

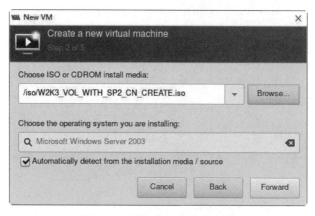

图 3-23　从本地 ISO 映像安装

第 3 步需要配置内存和 CPU 选项,这些值会影响宿主机和虚拟机的性能。向导提供默认的内存大小和 CPU 数量,我们也可以根据实际的需求进行调整,如图 3-24 所示。单击 Forward 按钮继续。

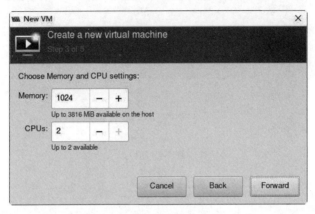

图 3-24　配置虚拟机的内存和 CPU

第 4 步需要配置虚拟机的存储,如图 3-25 所示。选中 Enable storage for this virtual machine 可以为新虚拟机启用存储,然后选中 Create a disk image for the virtual machine 单选按钮,会在宿主机上创建新磁盘映像文件,准确来讲就在那个名为 default 的存储池所对应的 /var/lib/libvirt/images/ 目录中创建新的映像文件。根据需要来指定文件的大小,本次实验保持默认磁盘映像的大小。单击 Forward 按钮。

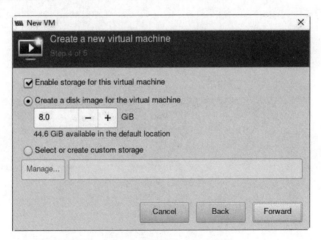

图 3-25　配置虚拟机的存储

第 5 步需要设置虚拟机的名称和检查最终配置,如图 3-26 所示。虚拟机名称可以包含字母、数字、下画线、句点和连字符,建议起一个有意义的虚拟机名称。默认情况下,新虚拟机将使用那个名为 default 的虚拟网络。

检查并验证一下虚拟机的设置，单击 Finish 按钮，这将创建具有指定网络设置、虚拟化类型和体系结构的虚拟机。

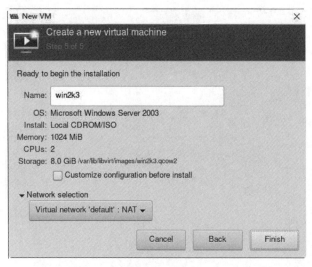

图 3-26　设置虚拟机的名称、网络并审核配置

在虚拟机中安装 Windows Server 2003 操作系统与物理机上安装过程类似，如图 3-27(a) 和图 3-27(b) 所示。

(a)

图 3-27　Windows Server 2003 虚拟机的安装

(b)

图 3-27 （续）

3.2.3 查看虚拟机与环境的配置

下面使用 virsh 命令查看虚拟机与环境的配置。查看当前虚拟机的列表，命令如下：

```
# virsh list
 Id    Name                           State
----------------------------------------------------
 1     centos6.10                     running
 3     win2k3                         running
```

查看虚拟机 w2k3 的信息，命令如下：

```
# virsh dominfo win2k3
Id:             3
Name:           win2k3
UUID:           b547eefa-37af-42af-97c2-730f9ac5288a
OS Type:        hvm
State:          running
CPU(s):         2
CPU time:       1276.9s
Max memory:     1048576 KiB
```

```
Used memory:         1048576 KiB
Persistent:          yes
Autostart:           disable
Managed save:        no
Security model:      seLinux
Security DOI:        0
Security label:      system_u:system_r:svirt_t:s0:c437,c725 (enforcing)
```

查看存储池 ISO 中的存储卷,命令如下:

```
# virsh vol-list iso
Name                 Path
-----------------------------------------------------------------
CentOS-6.10-i386-minimal.iso   /iso/CentOS-6.10-i386-minimal.iso
W2K3_VOL_WITH_SP2_CN_CREATE.iso /iso/W2K3_VOL_WITH_SP2_CN_CREATE.iso
```

查看存储池 default 中的存储卷,命令如下:

```
# virsh vol-list default
Name                 Path
-----------------------------------------------------------------
CentOS6.10.qcow2     /var/lib/libvirt/images/CentOS6.10.qcow2
win2k3.qcow2         /var/lib/libvirt/images/win2k3.qcow2
```

查看宿主机上网络连接的信息,命令如下:

```
# nmcli connection show
NAME    UUID                                  TYPE      DEVICE
ens32   0b1638a6-add5-4057-9bce-575efc3d5bf2  ethernet  ens32
virbr0  c847ab96-1add-47f2-b365-d2a882eab9f1  bridge    virbr0
vnet0   e606acdb-d7e4-475e-929c-f234a8164ffd  tun       vnet0
vnet1   8873aeef-12d4-4e79-a788-49df0fe17ccf  tun       vnet1
```

从输出的信息中可以看出来,新的网络连接名称很有规律,第 1 个虚拟机是 vnet0,第 2 个虚拟机是 vnet1,以此类推。

3.3 使用 virt-install 创建虚拟机

与前面两种方法相比,通过 virt-install 来创建新虚拟机是效率最高的方法,同时也是最复杂的方法。

通过 virt-install 来创建虚拟机需要满足两个先决条件:

1. 可访问的保存在本地或网络上的操作系统安装源

可以是以下其中之一:

(1) 安装介质的 ISO 映像文件。
(2) 现有操作系统的虚拟磁盘映像。

2．可实现的安装模式

(1) 如果是交互式安装,则需要 virt-viewer 软件。
(2) 如果是非交互式安装,就需要为操作系统安装程序提供回答文件,例如 kickstart 文件。

3.3.1 创建虚拟机并通过交互模式安装

使用 virt-install 创建虚拟机,必须提供以下参数。
(1) --name：虚拟机的名称。
(2) --memory：虚拟机的内存。
(3) --vcpus：虚拟机虚拟 CPU(vCPU)的数量。
(4) --disk：虚拟机磁盘类型和大小。
(5) 操作系统安装源的类型和位置,可以由--location、--cdrom、--pxe、--import 和--boot 选项来指定。

查看 virt-install 命令的帮助信息,命令如下：

```
#virt-install --help
```

查看 virt-install 选项属性的完整列表,命令如下：

```
#virt install --option=?
```

例如,查看磁盘存储和选项的命令如下：

```
#virt-install --disk=?
```

当然,最完整的帮助信息在 virt-install 手册页中,除了有每个命令选项之外,还有重要的提示和丰富的示例。

在运行 virt-install 之前,有可能还需要使用 qemu-img 命令来配置存储选项。有关 qemu-img 的使用,可参见后续的章节。

下面,我们使用 virt-install 命令在交互模式下安装一台 CentOS 6.10 的虚拟机。首先,要保证 virt-viewer 命令可以正常地启动。

```
#virt-viewer -V
virt-viewer version 7.0-9.el8 (OS ID: rhel8)
```

提示：如果系统中没有安装 virt-viewer,则可以通过命令 dnf -y install virt-viewer 进行安装。

下面准备创建了一个名为 centos6.10vm2 的虚拟机，给虚拟机分配了 1024MB 内存、1 个 vCPU、8GB 虚拟磁盘。将通过存储在本地 /iso/ 目录中的 CentOS-6.10-i386-minimal.iso 在虚拟机中安装操作系统。另外，通过 --os-variant 指定虚拟机操作系统的版本为 centos6.10。示例命令如下：

```
# virt-install -- name centos6.10vm2 \
-- memory 1048 -- vcpus 1 -- disk size = 8 \
-- cdrom /iso/centos-6.10-i386-minimal.iso \
-- os-variant centos6.10

Starting install...
Allocating 'centos6.10vm2.qcow2'           | 8.0 GB 00:00
```

由于仅指定虚拟磁盘的大小而未指定其存储位置和名称，所以 virt-install 会自动在默认的 default 存储池中创建一个名为 CentOS6.10vm2.qcow2 的存储卷，然后分配给虚拟机。

当虚拟创建完毕之后，virt-install 会自动启动 virt-viewer，然后就可以在其中通过交互模式进行安装了，如图 3-28 所示。

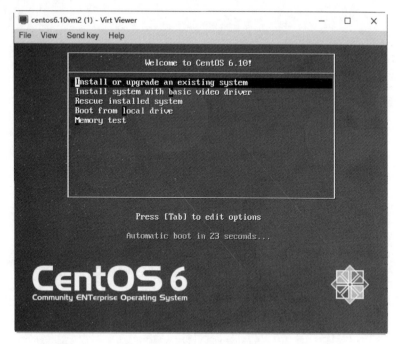

图 3-28　virt-viewer 是一个可与虚拟机交互的重要界面

3.3.2 查看虚拟机与环境的配置

安装完成之后,查看一下虚拟机的信息及宿主机存储的变化。

```
# virsh list
Id    Name                    State
----------------------------------------------------
4     centos6.10vm2           running

# virsh dominfo centos6.10vm2
Id:             4
Name:           centos6.10vm2
UUID:           e81c4a75-1d80-4c68-9951-bd16038b8bec
OS Type:        hvm
State:          running
CPU(s):         1
CPU time:       21.1s
Max memory:     1048576 KiB
Used memory:    1048576 KiB
Persistent:     yes
Autostart:      disable
Managed save:   no
Security model: seLinux
Security DOI:   0
Security label: system_u:system_r:svirt_t:s0:c206,c238 (enforcing)

# virsh domblklist CentOS6.10vm2
Target    Source
----------------------------------------------------
vda       /var/lib/libvirt/images/centos6.10vm2.qcow2
hda       -

# virsh vol-list default
Name                    Path
----------------------------------------------------------------------
centos6.10.qcow2        /var/lib/libvirt/images/centos6.10.qcow2
centos6.10vm2.qcow2     /var/lib/libvirt/images/centos6.10vm2.qcow2
win2k3.qcow2            /var/lib/libvirt/images/win2k3.qcow2
```

3.3.3 virt-install 高级用法示例

virt-install 有很丰富的选项参数以适用不同的场景,下面我们看两个示例。

示例 3-1:通过回答文件进行非交互式的自动化安装

在手工安装 RHEL/CentOS 时,需要设置多个选项参数,很烦琐。我们可以使用

Kickstart 文件实现自动化安装。Kickstart 文件是一个包含安装程序所需要的选项参数的文本文件。

默认情况下 RHEL/CentOS 的安装程序会在/root 目录生成一个名为 anaconda-ks.cfg 的文件,我们可以以这个文件的内容为"起点"快速生成一个回答文件,命令如下:

```
# cp anaconda-ks.cfg centos6.9.txt

# vi centos6.9.txt
    # 可以根据需求进行修改,以下参数适合于 CentOS 6
autostep
install
cdrom
lang en_US.UTF-8
keyboard us
text
network --onboot yes --device eth0 --bootproto dhcp --noipv6
rootpw 123456
#firewall --service=ssh
firewall --disabled
authconfig --enableshadow --passalgo=sha512
#seLinux --enforcing
seLinux --disabled
timezone --utc Asia/Shanghai
bootloader --location=mbr --driveorder=vda --append="crashKernel=auto rhgb quiet"
# The following is the partition information you requested
# Note that any partitions you deleted are not expressed
# here so unless you clear all partitions first, this is
# not guaranteed to work
#clearpart --all --drives=vda
#volgroup VolGroup --pesize=4096 pv.253002
#logvol / --fstype=ext4 --name=lv_root --vgname=VolGroup --grow --size=1024 --maxsize=51200
#logvol swap --name=lv_swap --vgname=VolGroup --grow --size=921 --maxsize=921
#part /boot --fstype=ext4 --size=500
#part pv.253002 --grow --size=1


#clearpart --all --initlabel --drives=vda
#part /boot --fstype=ext4 --size=200
#part pv.1 --grow --size 1
#volgroup vg0 pv.1
#logvol / --fstype=ext4 --name=lv_root --vgname=vg0 --grow
zerombr
clearpart --all --drives=vda
part /boot --fstype=ext4 --size=500
```

```
part pv.253002 --grow --size=1
volgroup VolGroup --pesize=4096 pv.253002
logvol swap --name=lv_swap --vgname=VolGroup --grow --size=921 --maxsize=921
logvol / --fstype=ext4 --name=lv_root --vgname=VolGroup --grow --size=1024 --
maxsize=51200

%packages --nobase
@core
%end
reboot
```

Kickstart 文件的参数设置可参见官方文档:https://access.redhat.com/documentation/en-us/red_hat_enterprise_Linux/6/html/installation_guide/s1-kickstart2-options。

除了这个回答文件,还需要为 virt-install 命令提供 3 个选项参数。

(1) --location:为 virt-install 指定 Linux 发行版本的安装源。既可以是网络位置,也可以是本地 ISO 文件,例如/iso/centos-6.10-i386-minimal.iso。

(2) --initrd-inject:需要与--location 选项一起使用,它指定一个保存在宿主机本地的 Kickstart 文件,例如/root/centos6.9.txt。

(3) --extra-args:需要与--location 选项一起使用,它传递给安装程序额外的参数,例如通过"ks=file:/centos6.9.txt"指定 Kickstart 文件。

示例命令如下:

```
# virt-install --name centos6.10vm3 \
--memory 1048 --vcpus 1 --disk size=8 \
--os-variant centos6.10 \
--location /iso/centos-6.10-i386-minimal.iso \
--initrd-inject /root/centos6.9.txt \
--extra-args="ks=file:/centos6.9.txt"

Starting install...
Retrieving file vmlinuz...                    | 4.0 MB 00:00
Retrieving file initrd.img...                 |  37 MB 00:00
Allocating 'centos6.10vm3.qcow2'              | 8.0 GB 00:00
PuTTY X11 proxy: unable to connect to forwarded X server: Network error: Connection refused
Unable to init server: Could not connect: Connection refused

(virt-viewer:2057): Gtk-WARNING **: 22:29:23.035: cannot open display: localhost:10.0
Domain installation still in progress. You can reconnect to
the console to complete the installation process.
```

当虚拟机创建完毕之后,上述这个 virt-install 命令并不会启动 virt-viewer。只要 Kickstart 文件内容正确,就会在非交互模式下进行全自动的安装。当然,也可以通过

Cockpit 或 virt-manager 中的控制台查看安装的过程。

示例 3-2：对虚拟机性能影响很大的 os-variant 选项

Cockpit 和 virt-manager 都会对安装介质进行检测并提供一个操作系统类型的值。虽然这个值不是必需的，但是强烈建议设置最接近的值，因为 libvirt 会根据它来对虚拟机进行有针对性的优化。

可以通过 --os-variant 选项为 virt-install 命令指定操作系统类型，例如 fedora32、rhel8、Windows 10。可以使用命令 osinfo-query 获取可接受的操作系统的列表。例如获得 CentOS 6、7、8 版本的字符串的命令如下：

```
#osinfo-query os | egrep centos[678]
centos6.0    | CentOS 6.0  | 6.0   | http://centos.org/centos/6.0
centos6.1    | CentOS 6.1  | 6.1   | http://centos.org/centos/6.1
centos6.10   | CentOS 6.10 | 6.10  | http://centos.org/centos/6.10
centos6.2    | CentOS 6.2  | 6.2   | http://centos.org/centos/6.2
centos6.3    | CentOS 6.3  | 6.3   | http://centos.org/centos/6.3
centos6.4    | CentOS 6.4  | 6.4   | http://centos.org/centos/6.4
centos6.5    | CentOS 6.5  | 6.5   | http://centos.org/centos/6.5
centos6.6    | CentOS 6.6  | 6.6   | http://centos.org/centos/6.6
centos6.7    | CentOS 6.7  | 6.7   | http://centos.org/centos/6.7
centos6.8    | CentOS 6.8  | 6.8   | http://centos.org/centos/6.8
centos6.9    | CentOS 6.9  | 6.9   | http://centos.org/centos/6.9
centos7.0    | CentOS 7    | 7     | http://centos.org/centos/7.0
centos8      | CentOS 8    | 8     | http://centos.org/centos/8
```

3.4 半虚拟化驱动 VirtIO

为了提高虚拟机的硬盘、网络及显卡设备的性能，需要在虚拟机中安装半虚拟化驱动程序 VirtIO 以替换普通的驱动程序。

3.4.1 半虚拟化驱动 VirtIO 原理

通过 virt-manager 可以查看虚拟机硬件的详细信息，例如虚拟磁盘总线接口，会看到的有 IDE、SATA、SCSI、USB 和 VirtIO 5 种类型，如图 3-29 所示。

类似地，虚拟机网卡有 e100、rtl8139 和 VirtIO 3 种类型，虚拟显卡有 Bochs、QXL、VGA 和 VirtIO 4 种类型。

上述 VirtIO 之外的虚拟设备均是全虚拟化（Full Virtualization）类型的设备。这种类型的设备的优点是适应性强，可以适合任何虚拟化操作系统，但是由于访问路径长，所以性能会比较差，如图 3-30 所示。

虚拟机操作系统为这些全虚拟化设备安装的是普通的驱动程序（通常在安装操作系统时会自动进行安装），它是不知道自己运行在虚拟机中。操作对设备的请求被 Hypervisor

拦截,然后转发给 QEMU,QEMU 翻译之后再转发给宿主机上的驱动程序,最后才到达真实的物理硬件设备。

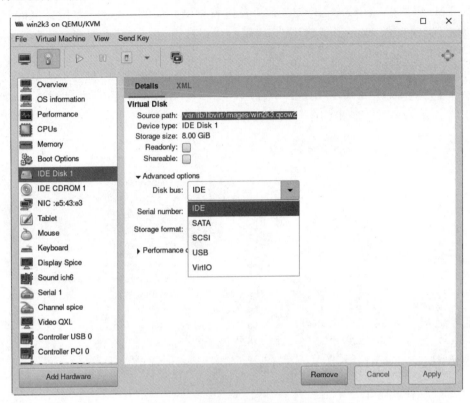

图 3-29 虚拟磁盘总线接口类型

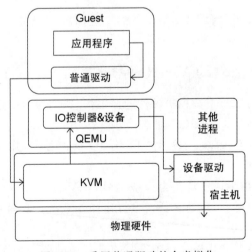

图 3-30 采用普通驱动的全虚拟化

如果为虚拟机分配的是 VirtIO 类型的虚拟设备,而且在虚拟机操作系统中安装了半虚拟化驱动 VirtIO,则这时虚拟机操作系统知道自己是虚拟机,所以数据直接发送给由 QEMU 提供的半虚拟化设备,而不经过 Hypervisor,通过宿主机上的驱动程序发送给真实的物理硬件设备,如图 3-31 所示。

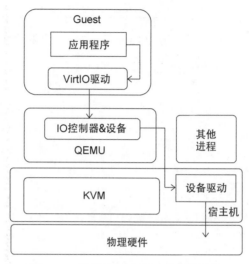

图 3-31 采用 VirtIO 驱动的半虚拟化

由于不与 Hypervisor 交互,所以半虚拟化驱动 VirtIO 缩短了访问路径,这种模式带来的好处就是性能要比采用普通的驱动好很多,从而让虚拟设备接近物理设备的性能。这个工作原理与 VMware Tools 驱动类似。

3.4.2 半虚拟化驱动 VirtIO 的安装

RHEL/CentOS 4.8、5.3 之后的发行版本都包含 VirtIO 设备的驱动程序,所以在虚拟机中安装这些发行版本时,如果安装程序检测到虚拟机的硬件是 VirtIO 类型的,就会自动安装相应的驱动程序。

由于 Windows 操作系统中不包含 VirtIO 设备的驱动程序,所以需要手工安装。在安装 Windows Server 2008 或之后的操作系统时,可以指定 VirtIO 驱动程序所在的位置(光盘、软件均可)。在安装 Windows Server 2003 或更早的操作系统时,安装程序仅会扫描并读取软盘中的驱动程序。

> 提示:微软公司于 2015 年 7 月就停止了对 Windows Server 2003 的支持,我们现在仅仅出于实验的目的使用它。

如何获得 VirtIO for Windows 驱动程序呢?一种方法是 RHEL/CentOS 8 的安装介质和软件仓库中包括 virtio-win 软件包,另外一种方法从 KVM 项目的网站下载。下载链接为 http://www.Linux-kvm.org/page/WindowsGuestDrivers/Download_Drivers。

下面，我们采用第 1 种方法来为 Windows Server 2019 的安装程序提供驱动程序。

提示：在嵌套环境中安装 Windows Server 2019 通常会很慢。如果有条件，则建议在物理环境做这个实验。

首先，检查一下宿主机上是否有 VirtIO for Windows 的驱动程序，命令如下：

```
# rpm -qi virtio-win
Name            : virtio-win
Version         : 1.9.12
Release         : 2.el8
Architecture: noarch
Install Date: Wed 05 Aug 2020 10:19:48 PM CST
Group           : Applications/System
Size            : 663922676
License         : Red Hat Proprietary and GPLv2
Signature       : RSA/SHA256, Wed 22 Jul 2020 03:23:43 AM CST, Key ID 05b555b38483c65d
Source RPM      : virtio-win-1.9.12-2.el8.src.rpm
Build Date      : Wed 22 Jul 2020 02:46:14 AM CST
Build Host      : aarch64-05.mbox.centos.org
Relocations     : (not relocatable)
Packager        : CentOS Buildsys <bugs@centos.org>
Vendor          : CentOS
URL             : http://www.redhat.com/
Summary         : VirtIO para-virtualized drivers for Windows(R)
Description :
VirtIO para-virtualized Windows(R) drivers for 32-bit and 64-bit Windows(R) guests.

# rpm -ql virtio-win | grep .iso
/usr/share/virtio-win/virtio-win-1.9.12.iso
/usr/share/virtio-win/virtio-win.iso

# ls -l /usr/share/virtio-win/*.iso
-rw-r--r--. 1 root root 332519424 Jul 22 02:46 /usr/share/virtio-win/virtio-win-1.9.
12.iso
lrwxrwxrwx. 1 root root 21 Jul 22 02:46 /usr/share/virtio-win/virtio-win.iso -> virtio-
win-1.9.12.iso
```

virtio-win.iso 是 virtio-win-1.9.12.iso 的符号链接文件。可以通过它为虚拟机提供驱动程序。

通过 virt-manager 创建 Windows Server 2019 的虚拟机。默认情况下，虚拟磁盘的接口为 SATA，我们将其修改为 VirtIO，如图 3-32 所示。

启动虚拟机，开始安装 Windows Server 2019。由于安装程序没有 VirtIO 设备的驱动程序，所以无法识别 VirtIO 接口的磁盘，它会提示"我们找不到任何驱动器。要获取存储设备驱动程序，请单击'加载驱动程序'"，如图 3-33 所示。

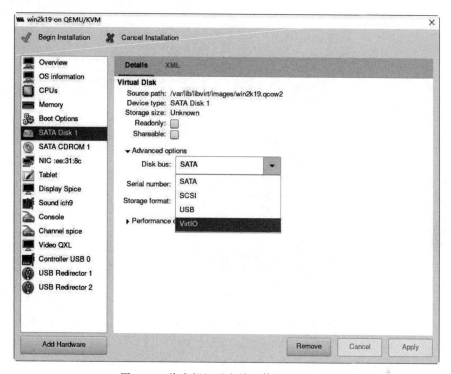

图 3-32　将虚拟机磁盘接口修改为 VirtIO

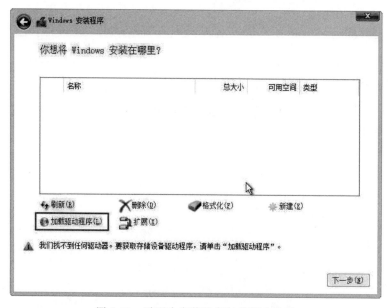

图 3-33　需要为安装程序加载驱动程序

在单击"加载驱动程序"之前,我们需要修改虚拟机的配置,为将虚拟光驱"换一张"光盘。在虚拟机配置中找到 CDROM,单击 Browse 按钮,如图 3-34 所示。

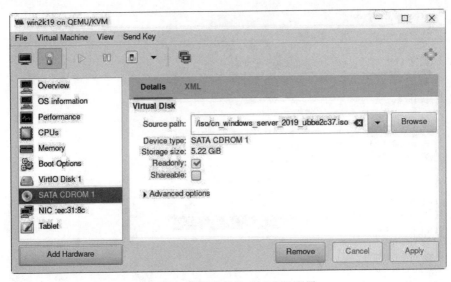

图 3-34　修改虚拟机 CDROM 配置

选中 /usr/share/virtio-win/ 目录中 virtio-win.iso 文件,单击 Open 按钮,如图 3-35 所示。

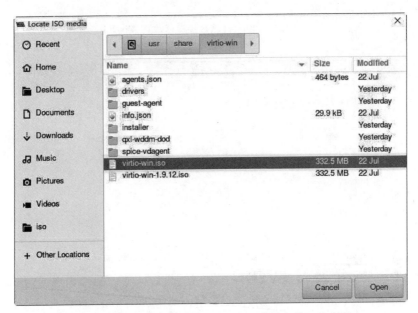

图 3-35　选中 VirtIO for Windows 驱动程序的 ISO 文件

这个操作相当于给虚拟机更换了一张光盘，单击 Apply 按钮，如图 3-36 所示。

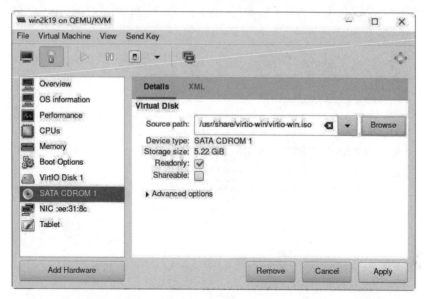

图 3-36　为虚拟机 CDROM 指定 VirtIO for Windows 驱动程序的 ISO 文件

返回 Windows 安装程序，单击"加载驱动程序"，会出现加载程序提示窗口。单击"确定"按钮，如图 3-37 所示。

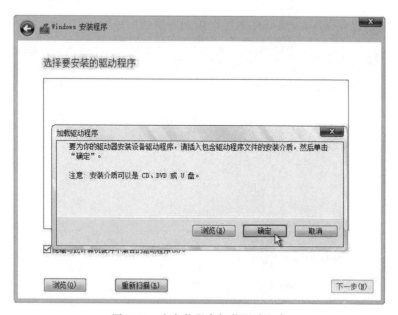

图 3-37　为安装程序加载驱动程序

安装程序会搜索 CD、DVD 或 U 盘，并将找到的驱动程序显示出来。在本实验中，选择 Windows Server 2019 的驱动程序(光盘中\amd64\2k19\目录下的 viostor.inf)，然后单击"下一步"按钮，如图 3-38 所示。

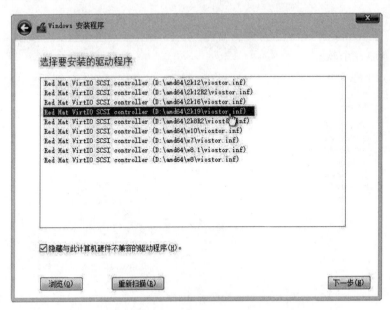

图 3-38　在找到的驱动程序中选择适合的版本

有了 VirtIO 驱动程序，安装程序就可以识别虚拟机的磁盘了，如图 3-39 所示。

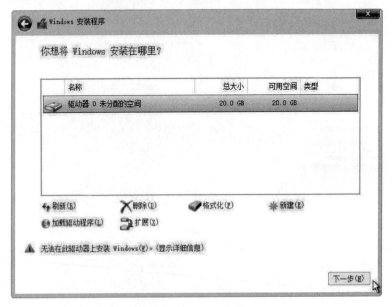

图 3-39　安装程序识别出 VirtIO 接口的磁盘

如果是 Windows Server 2003 或更早的操作系统，安装程序仅会扫描并读取保存在软盘中的驱动程序。最简单的方法是从 Fedora 社区下载包含 VirtIO 驱动程序的虚拟磁盘文件，其链接为 https://fedorapeople.org/groups/virt/virtio-win/direct-downloads/stable-virtio/virtio-win_x86.vfd。

通过 virt-manager 创建 Windows Server 2003 虚拟机，将虚拟磁盘的接口设置为 VirtIO，再添加一个软盘驱动器，在 Add New Virtual Hardware 中选中 Storage，然后在右框选中 Select or create custom storage，单击 Manage 按钮，如图 3-40 所示。

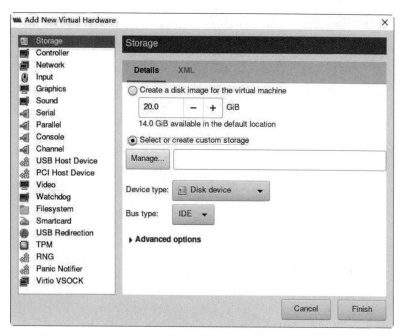

图 3-40　添加新的存储

指定包含 VirtIO 驱动程序的虚拟软盘文件 virtio-win_x86.vfd，如图 3-41 所示。

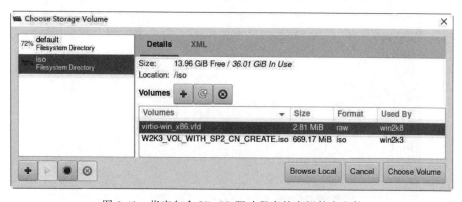

图 3-41　指定包含 VirtIO 驱动程序的虚拟软盘文件

然后将这个新存储设备的设备类型指定为 Floppy device，然后单击 Finish 按钮继续，如图 3-42 所示。

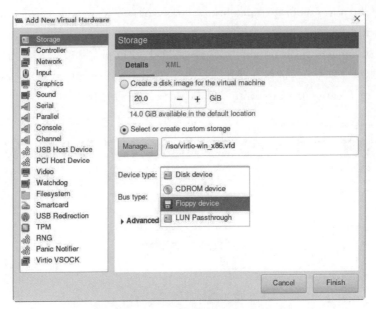

图 3-42　将新存储的设备类型指定为 Floppy device

Windows Server 2003 的安装程序在安装过程中会读取软盘。由于找到了驱动程序，所以会看到 VirtIO 接口的虚拟磁盘，如图 3-43 所示。

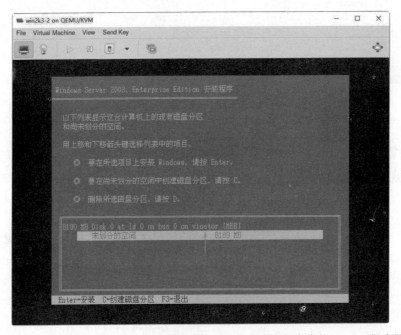

图 3-43　Windows Server 2003 安装程序读取并加载保存在软盘中的 VirtIO 驱动程序

不管通过哪种方式为 Windows 操作系统提供了 VirtIO 驱动程序，都会在设备管理器看到磁盘驱动器和存储控制器类型为 VirtIO SCSI 接口，如图 3-44 所示。

图 3-44　正确安装 VirtIO 存储控制器驱动程序的虚拟机

3.5　QEMU Guest Agent

3.5.1　QEMU Guest Agent 原理

为了更好地管理虚拟机，包括 KVM 在内的虚拟化平台都需要通过某种机制与虚拟机进行通信，这样既可以获得虚拟机操作系统的详细信息，也可以向虚拟机操作系统发出指令。例如：文件系统的冻结和解冻、系统挂起或安全地关闭等。VMware 虚拟机平台是通过在虚拟机中安装 VMware Tools 实现的，而 KVM 虚拟机则需要安装 QEMU Guest Agent 并正确运行，除此之外还需要为 KVM 虚拟机配置 VirtIO 串行控制器。

QEMU Guest Agent 是虚拟机上的守护程序或服务。想让它正常地运行，虚拟机上还必须有一个名为 Channel qemu-ga 的设备（ga 是 Guest Agent 的缩写），如图 3-45 所示。

宿主机上的 virsh、virt-manager 等基于 libvirt 的应用程序与虚拟机上的 QEMU Guest Agent 进行通信，此通信并不是通过网络进行的，而是通过虚拟机上 VirtIO 串行控制器设备进行的，如图 3-46 所示。这个 VirtIO 串口控制器设备是通过一个字符设备驱动程序（通常是 UNIX 套接字）连接到宿主机上的，而在虚拟机中 QEMU Guest Agent 负责侦听此串行通道上的信号。

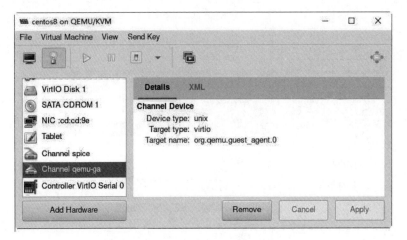

图 3-45　虚拟机上的 qemu-ga 通道设备

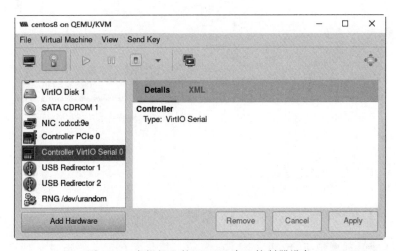

图 3-46　虚拟机上的 VirtIO 串口控制器设备

一旦有了 QEMU Guest Agent 的支持，很多基于 libvirt 的命令或应用程序功能变得更强大，以 virsh 命令为例：

（1）使用 virsh shutdown --mode = agent 命令来关闭虚拟机，这种关闭方法比命令 virsh shutdown --mode = acpi 更可靠，因为相当于在虚拟机操作系统中发出类似 shutdown 指令，从而确保以干净状态关闭虚拟机。

（2）使用 virsh domfsinfo 命令可以获得正在运行的虚拟机中已挂载文件系统的列表。

（3）使用 virsh domtime 命令可以查询或设置虚拟机的时钟。

（4）使用 virsh domifaddr --source agent 命令可以查询虚拟机操作系统的 IP 地址。

> 提示：宿主机通过 QEMU Guest Agent 与虚拟机交互的范围大小与功能的多寡还依赖于虚拟机的操作系统的版本。例如 Windows Server 2003 仅支持很少的命令。可以通过 virsh qemu-agent-command 虚拟机名称 '{"execute":"guest-info"}' 命令来查询支持的指令。

3.5.2　Linux 下的 QEMU Guest Agent

目前，不管是通过 Cockpit、virt-manager 还是 virt-install 创建的 Linux 虚拟机，默认都会配置 VirtIO 串口控制器和 qemu-ga 通道设备，而且 RHEL/CentOS 7 之后发行版本的安装程序，一旦检测出拥有 VirtIO 串口控制器的设备就会自动安装 QEMU Guest Agent，这使对 Linux 虚拟机的管理更加方便。命令如下：

```
# virsh domfsinfo centos8
Mountpoint     Name     Type     Target
-------------------------------------------
/              dm-0     ext4     vda
/boot          vda1     ext4     vda

# virsh reboot centos8 --mode agent
Domain centos8 is being rebooted
```

这两个命令之所以成功是由于虚拟机 centos8 有 VirtIO 串行控制器设备，而且 QEMU Guest Agent 守护程序正常运行。下面在虚拟机上进行查看，命令如下：

```
[root@guest ~]# cat /etc/system-release
CentOS Linux release 8.1.1911 (Core)

[root@guest ~]# systemctl status qemu-guest-agent.service
● qemu-guest-agent.service - QEMU Guest Agent
   Loaded: loaded (/usr/lib/systemd/system/qemu-guest-agent.service; enabled; v>
   Active: active (running) since Wed 2020-08-19 21:48:57 CST; 41min ago
 Main PID: 866 (qemu-ga)
    Tasks: 1 (limit: 11493)
   Memory: 2.1M
   CGroup: /system.slice/qemu-guest-agent.service
           └─866 /usr/bin/qemu-ga --method=virtio-serial --path=/dev/virtio-por>

Aug 19 21:48:57 localhost.localdomain systemd[1]: Started QEMU Guest Agent.
```

如果将虚拟机中的 QEMU Guest Agent 守护程序停止，命令如下：

```
[root@guest ~]# systemctl stop qemu-guest-agent.service
```

这时再在宿主机上使用 virsh 命令来与虚拟机进行交互，命令如下：

```
1 # virsh domfsinfo centos8
    error: Unable to get filesystem information
    error: Guest agent is not responding: QEMU guest agent is not connected

2 # virsh reboot centos8 -- mode agent
    error: Failed to reboot domain centos8
    error: Guest agent is not responding: QEMU guest agent is not connected

3 # virsh reboot centos8 -- mode acpi
    Domain centos8 is being rebooted
```

由于 QEMU Guest Agent 守护程序停止，所以第 1 和第 2 行命令执行失败。第 3 行命令成功的原因是指定的模式为 APCI（高级配置和电源管理接口 Advanced Configuration and Power Management Interface）。

如果是 RHEL/CentOS 6 的虚拟机，就需要手工来安装 QEMU Guest Agent，命令如下：

```
# cat /etc/system-release
  CentOS release 6.10 (Final)

# yum -y install qemu-guest-agent

# chkconfig qemu-ga on

# chkconfig -- list qemu-ga
  qemu-ga        0:off    1:off    2:on    3:on    4:on    5:on    6:off
```

提示：由于 CentOS 6 于 2020 年 11 月 30 日到期，原软件仓库已经被弃用，所以执行 YUM 命令会出错。解决方法是修改 /etc/yum.repos.d/CentOS-Base.repo，使用 vault 软件仓库源：

[base] 中的地址：baseURL=https://vault.centos.org/6.10/os/$basearch/
[updates] 中的地址：baseURL=https://vault.centos.org/6.10/updates/$basearch/
[extras] 中的地址：baseURL=https://vault.centos.org/6.10/extras/$basearch/

3.5.3 Windows 下的 QEMU Guest Agent

与创建 Linux 虚拟机不同，不管是通过 Cockpit、virt-manager 还是 virt-install 创建的

Windows 虚拟机,默认仅仅会配置 VirtIO 串口控制器设备,而没有 qemu-ga 通道设备,所以我们需要手工添加该设备。

在 Add New Virtual Hardware 窗口中选择 Channel,然后在 Name 的下拉列表框中选中 org.qemu.guest_agent.0,在 Device Type 中选择 Unix Socket(unix),选中 Auto socket,单击 Finish 按钮,如图 3-47 所示。

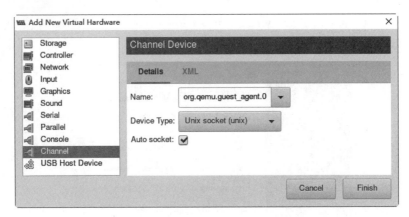

图 3-47　为虚拟机添加 qemu-ga 通道设备

有了 VirtIO 串口控制器和 qemu-ga 通道设备之后,还需要在 Windows 操作系统中安装 VirtIO 串口控制器的驱动程序及 qemu-ga 软件。如果没有安装驱动程序,则 Windows 虚拟机会将 VirtIO 串口控制器设备显示为"PCI 简单通信控制器",如图 3-48 所示。

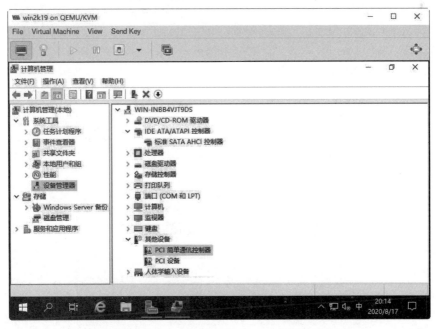

图 3-48　没有安装驱动程序的 VirtIO 串口控制器设备

> 提示：另外一个没有驱动程序的"PCI 设备"是 VirtIO Balloon 设备，这会在后续章节中介绍。

下面以 Windows Server 2019 的虚拟机为例，演示 VirtIO 串口控制器设备驱动及 QEMU Guest Agent 的安装。

首先，将宿主机的 /usr/share/virtio-win/virtio-win.iso 通过 CDROM 附加给虚拟机，然后在虚拟机中双击光盘中的 virtio-win-guest-tools.exe 来启动安装程序，如图 3-49 所示。

图 3-49　VirtIO-win guest tools 安装程序

VirtIO-win guest tools 除了包括多种 VirtIO 驱动程序之外，还包括 QEMU Guest Agent、QEMU Guest Agent VSS Provider 和 Spice Agent 共 3 个代理服务，如图 3-50 所示，保持默认值，单击 Next 按钮开始安装。

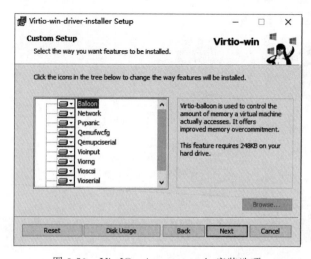

图 3-50　VirtIO-win guest tools 安装选项

安装结束之后，会在设备管理器中看到安装了正确驱动程序的 VirtIO 设备，如图 3-51 所示。

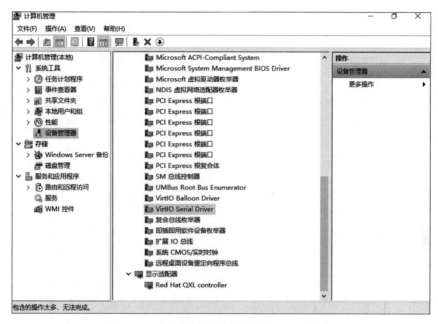

图 3-51　安装了正确驱动程序的 VirtIO 设备

在系统服务中会看到新增的 QEMU Guest Agent、QEMU Guest Agent VSS Provider 和 Spice Agent 等服务，如图 3-52 所示。

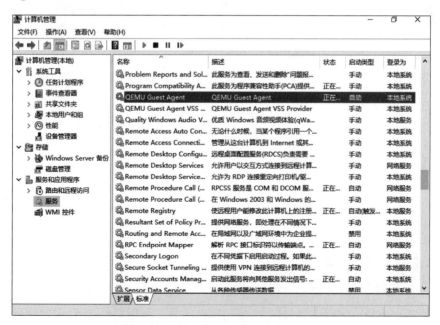

图 3-52　QEMU Guest 的 3 个代理服务

下面,我们在宿主机上通过 virsh 命令与一个名为 win2k19 的虚拟机进行交互操作,包括重置 administrator 的密码,命令如下:

```
# virsh domfsinfo win2k19
Mountpoint        Name              Type        Target
----------------------------------------------------------------
System Reserved   \\?\Volume{6ecd2dc1-0000-0000-0000-100000000000}\  NTFS
C:\               \\?\Volume{6ecd2dc1-0000-0000-0000-602200000000}\  NTFS
D:\               \\?\Volume{f6e0b9f5-df06-11ea-a74a-806e6f6e6963}\  UDF
E:\               \\?\Volume{f6e0b9f6-df06-11ea-a74a-806e6f6e6963}\  CDFS

# virsh domifaddr win2k19 --source agent
Name              MAC address         Protocol     Address
----------------------------------------------------------------
以太网             52:54:00:9c:85:e5   ipv6         fe80::c50:cad9:2ddf:c8f8%6/64
-                 -                   ipv4         192.168.122.94/24
Loopback Pseudo-Interface 1           ipv6         ::1/128
-                 -                   ipv4         127.0.0.1/8

# virsh set-user-password win2k19 --user administrator --password P@ssw0rd1
Password set successfully for administrator in win2k19
```

宿主机的 /usr/share/virtio-win/virtio-win.iso 其实是同目录下 virtio-win-1.9.12.iso 的符号链接。这个版本的 virtio-win-guest-tools.exe 仅支持 Windows 8、Windows Server 2012 及以后的 Windows 的版本,即使手工安装驱动及软件也仅仅可支持到 Windows 7。

如果需要使用更早的 Windows 操作系统,例如在实验环境中需要使用 Windows Server 2003,则可以在 Fedora 社区网站上下载早期的 VirtIO 驱动和 QEMU Guest Agent 软件,下载网址为 https://fedorapeople.org/groups/virt/virtio-win/direct-downloads。

提示:笔者在 Windows Server 2003 下测试过 virtio-win-0.1.100.iso、virtio-win-0.1.118.iso,均可正常运行。

3.6 显示设备与协议

为了让虚拟机显示图形,需要为其提供两个组件:显示设备(显卡)及从客户端访问图形的方法或协议。

3.6.1 显示设备

显示设备(显卡)是计算机的输出设备,但并不是必需的设备,例如:对于 Linux 来讲,如果不需要显示图形或进行控制台操作,串口的输出就工作得很好,但是多数情况下还是会为虚拟机配置显示设备。

虽然可以将宿主机的显卡通过透传的方式分配给虚拟机,但是最常见的还是为其分配虚拟显卡。

libvirt 官方目前支持 vga、cirrus、vmvga、xen、vbox、qxl、virtio、gop、bochs、ramfb、none 等 11 种不同类型的虚拟/仿真图形适配器,它们的区别主要是功能和分辨率。目前在 Cockpit 中无法配置虚拟显卡的类型,只能使用默认的 QXL。virt-manager 中支持 4 种虚拟显卡,默认也是 QXL,如图 3-53 所示。

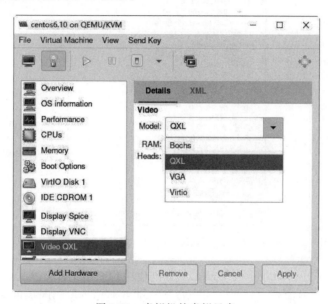

图 3-53 虚拟机的虚拟显卡

QXL 是具有 2D 支持的半虚拟图形驱动程序,其默认的 VGA 内存大小为 16MB,可以满足 2560×1440 分辨率需求,而且是与 SPICE 远程显示协议配合得最好的虚拟图形卡,所以推荐使用 QXL 虚拟图形适配器。

在使用 QXL 虚拟图形适配器时,通常保持默认的配置参数就可以很好地工作了。

```
<video>
  <model type='qxl'/>
</video>
```

对于 Windows 操作系统来讲,还需要安装 QXL 显卡驱动程序(可以在 virtio-win 软件包中找到)。安装了 QXL 虚拟显卡驱动程序的 Windows 操作系统,如图 3-54 所示。

3.6.2 显示协议

客户端访问远程主机图形界面的方法或协议有很多种,包括著名的 RDP、VNC、

SPICE、PCoIP 和 HDX，还有"小众"的 CodeCraft、NoMachine、FreeNX 等。它们之间的区别主要是色彩深度、加密、声频、带宽管理、文件系统重定向、打印机重定向、USB 和其他端口重定向等。本书仅涉及开源世界中最常见的 2 个协议 VNC 和 SPICE。

图 3-54　安装了 QXL 虚拟显卡驱动程序的 Windows Server 2003

通过 Cockpit、virt-manager 或 virt-install 创建虚拟机时，默认情况下会自动为其创建图形设备。图形设备对应的协议是显示协议。目前有两种类型的图形设备。

（1）SPICE：SPICE 是 Simple Protocol for Independent Computing Environment 协议的缩写。从本质上来讲，它仅仅是一种通信通道，它对在通道传递的数据没有要求。在 KVM 环境中，可以在这个通道中传递图像、声音及 USB 设备流等多种数据。

（2）VNC：VNC 是 Virtual Network Console 协议的缩写。VNC 是一种远程控制协议，可以实现对虚拟机的远程控制。

可以同时把这两种图形设备添加给虚拟机，如图 3-55 所示。当然，如果不需要通过控制台访问虚拟机，则可以不给虚拟机配置任何图形设备。

这两种图形设备都是 C/S 架构的。如果为虚拟机配置了 SPICE 图形设备，它就是 SPICE 的服务器端，常用的客户端有 virt-manager、virt-viewer。如果为虚拟机配置了 VNC 图形设备，它就是 VNC 的服务器端，除了 virt-manager、virt-viewer 之外，传统的 VNC 客户端如 VNC Viewer、noVNC 都可以访问虚拟机的控制台。

提示：noVNC 提供了一种使用支持 HTML5 Canvas 的浏览器访问 VNC Server 的方法。包括 Cockpit 在内的很多云计算、虚拟机管理软件采用了 noVNC 访问控制台。

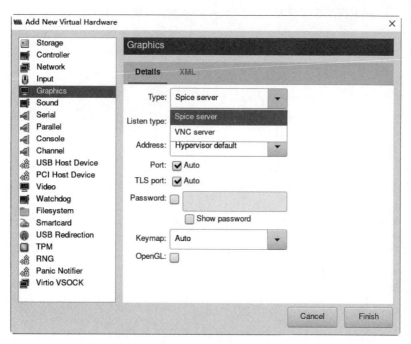

图 3-55 为虚拟机添加图形显示设备

出于安全考虑，QEMU 配置文件中的 SPICE 和 VNC 的默认监听地址为 127.0.0.1，也就是仅允许来自宿主机自身的 SPICE 和 VNC 客户端进行连接。查看 QEMU 配置的命令如下：

```
# cat /etc/libvirt/qemu.conf
…
# VNC is configured to listen on 127.0.0.1 by default.
# To make it listen on all public interfaces, uncomment
# this next option.
#
# NB, strong recommendation to enable TLS + x509 certificate
# verification when allowing public access
#
# vnc_listen = "0.0.0.0"
…
# SPICE is configured to listen on 127.0.0.1 by default.
# To make it listen on all public interfaces, uncomment
# this next option.
#
# NB, strong recommendation to enable TLS + x509 certificate
# verification when allowing public access
#
# spice_listen = "0.0.0.0"
…
```

默认情况下,新创建的虚拟机都会"继承"这些默认的配置。如果虚拟机未启动,则可在 virt-manager 中查看 VNC 和 SPICE 监听地址,显示的配置为 Hypervisor Default,如图 3-56(a)所示。

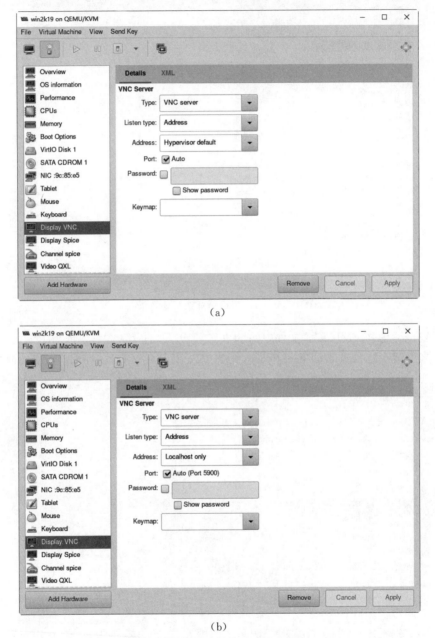

图 3-56 虚拟机图形显示设备配置

启动虚拟机之后,再在 virt-manager 中查看 VNC 和 SPICE 监听地址,会发现配置变为 Localhost only,同时还会显示自动生成的监听端口,例如 5900,如图 3-56(b)所示。

可以在宿主机的命令行中查看监听的地址与端口,命令如下:

```
#netstat -anp|grep 590
tcp    0    0 127.0.0.1:5900 0.0.0.0:*   LISTEN 18408/qemu-kvm
tcp    0    0 127.0.0.1:5901 0.0.0.0:*   LISTEN 18408/qemu-kvm
```

在本示例中,VNC 型的图形设备监听地址是 127.0.0.1、监听端口是 5900,SPICE 型的图形设备监听地址也是 127.0.0.1,不过监听端口则是 5901。

也可以在 Cockpit 虚拟机控制台中查看监听的地址与端口,如图 3-57 所示。

图 3-57　Cockpit 虚拟机控制台中显示的图形设备监听地址与端口

3.6.3　Remote Viewer 连接虚拟机排错

掌握了 KVM 图形设备的原理之后,下面我们来分析一个故障案例。

我们在一台 Windows 主机上通过 Google Chrome 或 Firefox 等浏览器远程访问宿主机上的 Cockpit。

如果给虚拟机配置了 VNC 型的图形设备,则可以在 Cockpit 虚拟机台中使用"图形控制台(VNC)"来正常地访问虚拟机,如图 3-58 所示。

这是由于 Cockpit 的 noVNC 运行在宿主机上,所以它使用 127.0.0.1 访问虚拟机当然没有问题。

1. 故障现象

如果给虚拟机配置了 SPICE 型的图形设备,则可以在 Cockpit 虚拟机控制台中选择"Desktop Viewer 中的图形控制台",单击"加载 Remote Viewer"按钮,浏览器会下载一个以 vv 为后缀的文件(vv 是 virt-viewer 的缩写),如图 3-59 所示。

如果这台 Windows 主机已经安装了 Remote Viewer 软件(Windows 下的 virt-viewer),由于文档关联的原因,会自动启动 Remote Viewer 软件,但是几秒之后,会提示"无法连接到图形服务器",如图 3-60 所示。

图 3-58　在 Cockpit 虚拟机控制台中使用 VNC 访问虚拟机

图 3-59　在 Cockpit 虚拟机控制台中使用 Remote Viewer 访问虚拟机

图 3-60　Remote Viewer 软件提示无法连接到图形服务器

单击"确定"按钮,Remote Viewer 软件会自动退出,而且下载的那个以 vv 为后缀的文件也不见了。

2. 分析排错

以 vv 为后缀的文件其实是一个文本文件。我们在出现错误提示时,如果不单击"确定"按钮,保持提示窗口不关闭,则可找到下载的以 vv 为后缀的文件,通过文本编辑器打开它,会看到其内容如下:

```
[virt-viewer]
type=spice
host=127.0.0.1
port=5901
delete-this-file=1
fullscreen=0

[..............................GraphicsConsole]
```

这个文件中 host 和 port 所指定的 IP 地址和端口就是 Remote Viewer 软件要访问的目标。也就是说 Remote Viewer 访问的是 127.0.0.1,而不是宿主机的 IP 地址。

故障原因找到了,下面我们就进行排错操作。在实验环境中,最简单的方法就是将虚拟机的两种图形设备的监听地址都修改为宿主机的 IP 地址,如图 3-61 所示。

重新启用虚拟机之后,Cockpit 控制台中显示的监听地址就变成了宿主机的 IP 地址,如图 3-62 所示。

在宿主机的命令行中查看监听的地址与端口,命令如下:

```
#netstat -an | grep 590
tcp    0    0 192.168.114.231:5900    0.0.0.0:*    LISTEN
tcp    0    0 192.168.114.231:5901    0.0.0.0:*    LISTEN
```

在 Cockpit 控制台中单击"加载 Remote Viewer"按钮,浏览器便会下载配置文件,这次使用 Remote Viewer 就可以正常地访问虚拟机了,如图 3-63 所示。

这次下载的以 vv 为后缀的文件的内容如下:

```
[virt-viewer]
type=spice
```

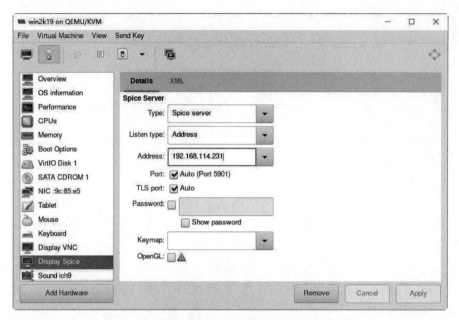

图 3-61　修改虚拟机图形显示设备监听的地址

图 3-62　Cockpit 虚拟机控制台中显示的图形设备监听的地址与端口

```
host = 192.168.114.231
port = 5901
delete-this-file = 1
fullscreen = 0


[...............................GraphicsConsole]
```

图 3-63　Remote Viewer 软件连接到虚拟机控制台

当然还要考虑宿主机上的防火墙配置。由于虚拟机图形设备监听的端口是 5900＋序号，所以根据宿主机上同时运行虚拟机的数量可以指定一个端口范围。例如端口范围是 5900～5950，命令如下：

```
# firewall-cmd --add-port=5900-5950/tcp --permanent

# firewall-cmd --add-port=5900-5950/tcp

# firewall-cmd --list-all
public (active)
  target: default
  icmp-block-inversion: no
  interfaces: ens32
  sources:
  services: cockpit dhcpv6-client rdp ssh vnc-server
  ports: 5900-5950/tcp
  protocols:
  masquerade: no
  forward-ports:
  source-ports:
  icmp-blocks:
  rich rules:
```

3.6.4 Linux 下的 SPICE Agent

与 QEMU Guest Agent 类似，SPICE Agent 也是一种运行在 KVM 虚拟机中的守护程序或服务软件，它可以与 SPICE 客户端软件（例如：virt-manager、virt-viewer）协助工作，为用户提供更流畅的图形显示。例如，在 virt-manager 中调整窗口大小时，SPICE Agent 会自动调整虚拟机的分辨率来适应窗口的大小。除此之外，现在 SPICE Agent 还支持宿主机和虚拟机之间剪贴板数据的交换、USB 设备的重定向等功能。

在虚拟机上安装 SPICE Agent 之前，需要保证虚拟机上有 spice 通道设备，如图 3-64 所示。

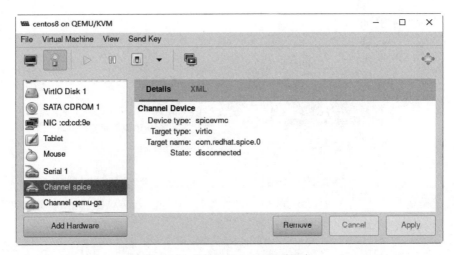

图 3-64 虚拟机的 spice 通道设备

然后，在 Linux 虚拟机中通过 yum 或 dnf 命令安装 SPICE Agent，命令如下：

```
[root@guest ~]# cat /etc/system-release
CentOS Linux release 8.3.2011

[root@guest ~]# dnf -y install spice-vdagent

[root@guest ~]# rpm -qi spice-vdagent
Name        : spice-vdagent
Version     : 0.20.0
Release     : 1.el8
...
URL         : https://spice-space.org/
Summary     : Agent for Spice guests
Description :
Spice agent for Linux guests offering the following features:
```

```
Features:
 * Client mouse mode (no need to grab mouse by client, no mouse lag)
   this is handled by the daemon by feeding mouse events into the Kernel
   via uinput. This will only work if the active X-session is running a
   spice-vdagent process so that its resolution can be determined.
 * Automatic adjustment of the X-session resolution to the client resolution
 * Support of copy and paste (text and images) between the active X-session and the client

[root@guest ~]# systemctl start spice-vdagentd
```

提示：在虚拟机中安装 SPICE Agent 的目的是为了提高图形显示效果，所以如果 Linux 虚拟机中不使用 X Windows，则可以不安装 SPICE Agent。

3.6.5　Windows 下的 SPICE Agent

如果希望提高 Windows 虚拟机的图形界面的显示效果，则强烈建议安装 SPICE Agent。

将 VirtIO-win 的 ISO 文件分配给虚拟机，双击 virtio-win-guest-tools.exe 安装程序。安装程序会自动安装 spice-guest-agent，如图 3-65 所示。

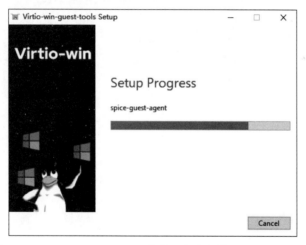

图 3-65　virtio-win-guest-tools 安装程序会安装 spice-guest-agent

安装之后，会看到系统中有一个名为 Spice Agent 的系统服务，如图 3-66 所示。

在这个服务的辅助下，当调整 virt-viewer 窗口的大小时，虚拟机图像也会随之进行缩放，同时还可以在宿主机与虚拟机之间通过剪贴板进行文本内容的交换。

图 3-66 Spice Agent 服务

3.7 本章小结

本章讲解了如何通过 Cockpit、virt-manager 和 virt-install 来安装虚拟机，VirtIO 驱动程序、QEMU Guest Agent、显示设备与协议原理、SPICE Agent 的工作原理及安装。

第 4 章将讲解虚拟机的管理。

第 4 章 管理虚拟机

在 RHEL/CentOS 8 中,最容易掌握的管理虚拟机的工具是图形界面的 Cockpit 和 virt-manager,但效率最高、功能最强大的则是命令行管理工具 virsh。

本章要点:
- libvirt 架构概述。
- 使用 virt-manager 管理虚拟机。
- 使用 virsh 管理虚拟机。
- 使用 Cockpit 管理虚拟机。

4.1 libvirt 架构概述

在本书的前三章,我们经常提及 libvirt,知道它是一个管理虚拟机及其他虚拟化功能的软件的集合,包括 API 库、守护进程(libvirtd)和其他一些工具。KVM 将 libvirt 作为虚拟化管理的引擎,而 Cockpit、virt-manager 及 virsh 等管理工具都是通过 libvirtd 进行管理操作的,它们将请求发送给 libvirtd,libvirtd 根据配置文件对虚拟化平台的计算、存储、网络等资源进行管理,如图 4-1 所示。

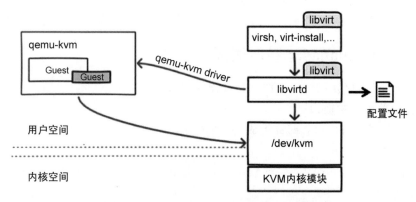

图 4-1　libvirt 是 KVM 虚拟化管理的引擎

在 RHEL/CentOS 中，要想正常地管理虚拟化平台，就一定要保证 libvirtd 守护程序处于正常运行的状态。下面我们通过实验来验证一下：

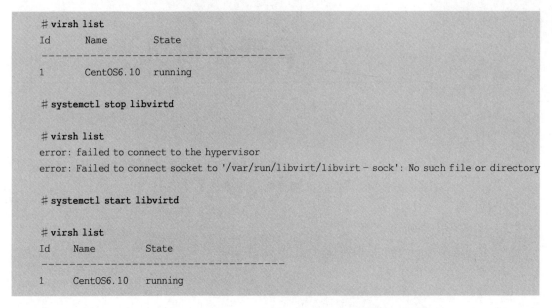

```
# virsh list
 Id    Name                           State
----------------------------------------------------
 1     CentOS6.10                     running

# systemctl stop libvirtd

# virsh list
error: failed to connect to the hypervisor
error: Failed to connect socket to '/var/run/libvirt/libvirt-sock': No such file or directory

# systemctl start libvirtd

# virsh list
 Id    Name                           State
----------------------------------------------------
 1     CentOS6.10                     running
```

从上述这个实验可以看出：如果 libvirtd 处于停止状态时，则连最基本的 virsh list 命令都会执行失败。

4.2 使用 virt-manager 管理虚拟机

virt-manager 是一种通过 libvirt 来管理虚拟化平台的 GUI 应用程序。我们可以通过它的向导功能来创建虚拟机，配置和调整虚拟机的资源设置，对正在运行的虚拟机和宿主机进行监控，通过内置的 VNC 和 SPICE 客户端访问虚拟机的控制台。另外，还可以通过它对远程宿主机进行管理。

4.2.1 virt-manager 界面概述

virt-manager 用户界面包括虚拟机管理器主窗口和虚拟机窗口。

1. 虚拟机管理器主窗口

虚拟机管理器主窗口由 3 部分组成，分别是菜单栏、工具栏和虚拟机列表窗格，如图 4-2 所示。

菜单栏包括 File、Edit、View 和 Help 共 4 个菜单，其中 File 菜单有 4 个子菜单，如表 4-1 所示。

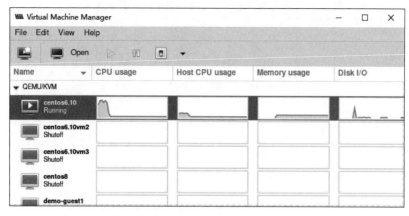

图 4-2　virt-manager 的虚拟机管理器主窗口

表 4-1　虚拟机管理器的 File 菜单

子菜单	说明
Add Connection	打开 Add Connection 对话框以连接到本地或远程的 Hypervisor
New Virtual Machine	打开 New VM 向导以创建新的虚拟机
Close	关闭虚拟机管理器主窗口,但不关闭任何虚拟机窗口,正在运行的虚拟机不会停止
Exit	关闭虚拟机管理器主窗口和所有的虚拟机窗口,正在运行的虚拟机不会停止

Edit 菜单也有 4 个子菜单,如表 4-2 所示。

表 4-2　虚拟机管理器的 Edit 菜单

子菜单	说明
Connection Details	打开选定连接的 Connection Details 窗口
Virtual Machine Details	打开选定虚拟机的 Virtual Machine 窗口
Delete	删除选定的连接或虚拟机
Preferences	打开用于配置虚拟机管理器选项的 Preferences 对话框

View 菜单仅有一个 Graphic 子菜单,可以控制在虚拟机管理器主窗口显示的状态信息,包括:

(1) Guest CPU Usage:虚拟机 CPU 的使用情况。

(2) Host CPU Usage:宿主机 CPU 的使用情况。

(3) Memory Usage:内存的使用情况。

(4) Disk I/O:磁盘 I/O 状态。

(5) Network I/O:网络 I/O 状态。

Preferences 子菜单中的 Polling 选项卡中的设置可以控制上述显示内容。

Help 菜单显示虚拟机管理器的版本信息。

虚拟机管理器主窗口工具栏中有 6 个图标,其功能如表 4-3 所示。

表 4-3　虚拟机管理器主窗口工具栏中的图标及其功能

图标	功能
	打开 New VM 向导以创建新的虚拟机
Open	打开所选虚拟机的虚拟机窗口
▶	启动选定的虚拟机
‖	暂停选定的虚拟机
■	停止选定的虚拟机
▼	打开子菜单以对所选虚拟机执行以下操作 （1）Reboot：重新启动选定的虚拟机 （2）Shut Down：关闭选定的虚拟机 （3）Force Reset：强制关闭选定的虚拟机并重新启动 （4）Force Off：强制关闭选定的虚拟机 （5）Save：将所选虚拟机的状态保存到文件中

2．虚拟机窗口

在虚拟机管理器主窗口中，双击虚拟机列表中的虚拟机或单击 Open 图标按钮就可以打开虚拟机窗口了。虚拟机窗口由 3 部分组成，分别是菜单栏、工具栏和虚拟机控制台窗格，如图 4-3 所示。

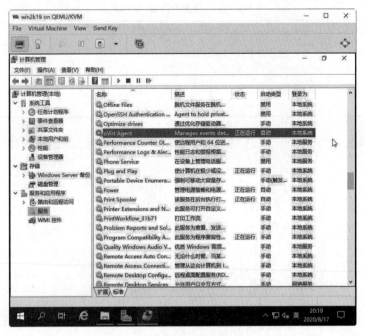

图 4-3　virt-manager 的虚拟机窗口

菜单栏包括 File、Virtual Machine、View 和 Send Key 共 4 个菜单,其中 File 菜单有 3 个子菜单,如表 4-4 所示。

表 4-4 虚拟机窗口中的 File 菜单

子菜单	说明
View Manager	打开虚拟机管理器主窗口
Close	仅关闭虚拟机窗口而不停止虚拟机
Quit	关闭所有"虚拟机管理器"窗口,正在运行的虚拟机不会停止

Virtual Machine 菜单共有 8 个子菜单,如表 4-5 所示。

表 4-5 虚拟机窗口中的 Virtual Machine 菜单

子菜单	说明
Run	运行虚拟机。仅当虚拟机未运行时,此选项才可用
Pause	暂停虚拟机。仅当虚拟机已在运行时,此选项才可用
Shut Down	打开子菜单,对此虚拟机上执行以下操作之一 (1) Reboot:重新启动虚拟机 (2) Shut Down:关闭虚拟机 (3) Force Reset:强制关闭虚拟机并重新启动 (4) Force Off:强制关闭虚拟机 (5) Save:将虚拟机的状态保存到文件中
Clone	创建虚拟机的副本
Migrate	打开 Migrate the virtual machine 对话框,以将虚拟机迁移到其他宿主机
Delete	删除虚拟机
Take Screenshot	拍摄虚拟机控制台的屏幕快照
Redirect USB Device	打开 Select USB devices for redirection 对话框,以选择要重定向的 USB 设备

View 菜单共有 8 个子菜单,如表 4-6 所示。

表 4-6 虚拟机窗口中的 View 菜单

子菜单	说明
Console	在下方的虚拟机窗格中打开控制台
Details	在下方的虚拟机窗格中查看详细信息
Snapshots	在下方的虚拟机窗格中打开快照
Fullscreen	以全屏模式显示虚拟机控制台
Resize to VM	设置为全屏显示并调整虚拟机的分辨率
Scale Display	基于以下子菜单项的选择对虚拟机的显示大小进行缩放 (1) Always:虚拟机的显示始终根据虚拟机窗格的大小进行缩放 (2) Only when Fullscreen:仅在虚拟机窗口处于全屏模式时,虚拟机的显示才会缩放到虚拟机窗口 (3) Never:虚拟机的显示永远不会缩放到虚拟机窗口 (4) Auto resize VM with window:调整虚拟机窗口大小时,虚拟机的显示会自动调整大小
Text Consoles	根据列表的值显示虚拟机,可选的值有 Serial 1 和 Graphical Console Spice
Toolbar	切换虚拟机窗口工具栏的显示或关闭

使用 Send Key 菜单中的设置值，可以将选定的快捷键向虚拟机传送，例如 Windows 操作系统常用的快捷键是 Ctrl＋Alt＋Delete。

虚拟机窗口的工具栏中共有 8 个图标，其功能如表 4-7 所示。

表 4-7 虚拟机窗口工具栏中的图标及其功能

图标	功　　能
 🖥	显示虚拟机的图形控制台
 💡	显示虚拟机的详细信息窗格
 ▷	启动虚拟机
 ❚❚	暂停虚拟机
 ◻	停止虚拟机
 ▼	打开子菜单以对所选虚拟机执行以下操作之一 (1) Reboot：重新启动虚拟机 (2) Shut Down：关闭虚拟机 (3) Force Reset：强制关闭虚拟机并重新启动 (4) Force Off：强制关闭虚拟机 (5) Save：将所选虚拟机的状态保存到文件中
 🖼	打开虚拟机快照
 ✥	以全屏模式显示虚拟机控制台

在 View 菜中单击 Details 或单击工具栏中的 💡 图标，就会显示虚拟机详细信息窗口，如图 4-4 所示。

虚拟机详细信息窗口包括虚拟机参数列表。选择列表中的参数后，与此参数有关的信息就会显示在虚拟机详细信息窗口的右侧。我们还可以在虚拟机详细信息窗口中添加、删除和配置虚拟硬件。

在 View 菜中单击 Snapshots 或单击工具栏中的 🖼 图标就会显示虚拟机快照窗口，如图 4-5 所示。

虚拟机快照窗口包括虚拟机的快照列表。选择列表中的快照后，与此快照有关的信息就会显示在窗口的右侧，包括状态、描述和屏幕快照等信息。我们可以在此处添加、删除和运行快照。

第4章 管理虚拟机 121

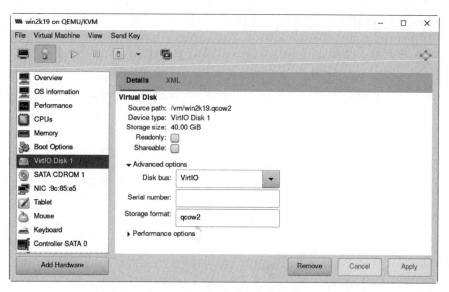

图 4-4　虚拟机的详细信息窗口

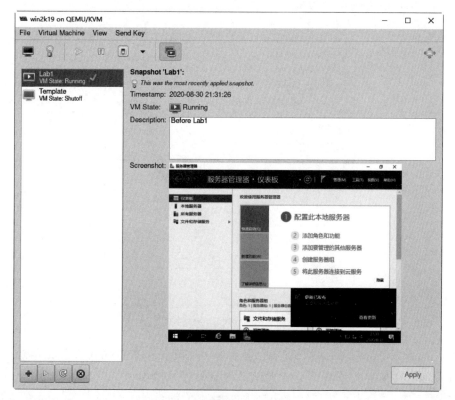

图 4-5　虚拟机的快照窗口

4.2.2 虚拟机生命周期管理

虚拟机的状态共有 7 种,如表 4-8 所示。其中最常见的是 shut off 和 running。

表 4-8 虚拟机的状态

状态	说明
running	虚拟机正在运行
idle	虚拟机处于空闲状态,即未运行或不可运行状态
paused	虚拟机已被暂停。通常是由管理员在 virt-manager 中进行 pause 操作或执行 virsh suspend 命令而发生的。当处于暂停状态时,虚拟机仍将消耗已分配的资源(例如内存)
in shutdown	当虚拟机操作系统收到关闭通知时,处于正常关闭过程中的状态
shut off	虚拟机未运行。通常表示虚拟机已被完全关闭或尚未启动
crashed	虚拟机已崩溃。通常是暴力关闭的结果,只有将虚拟机配置为在崩溃时不重新启动时,才会出现此状态
pmsuspended	虚拟机被虚拟机操作系统的电源管理软件设置为处于暂停状态

对处于 shut off 状态的虚拟机来讲,使用 virt-manager 来改变其状态的操作只有 Run,如图 4-6 所示。

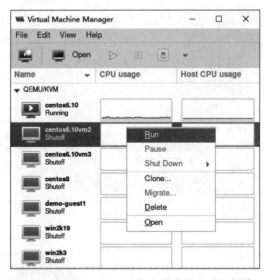

图 4-6 对处于 shut off 状态的虚拟机进行操作

对处于 running 状态的虚拟机来讲,使用 virt-manager 来改变其状态的操作有 Pause 和 Shut Down 两种,同时 Shut Down 的方式又细分为 5 种,如图 4-7 所示。

当单击 Pause 时,虚拟机会保持在当前状态,虚拟机的内存数据还会保留在宿主机的内存中,但是会释放所占用的处理器资源,也不会发生磁盘和网络 I/O 操作。这时,虚拟机的

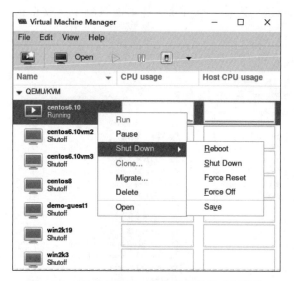

图 4-7 对处于 running 状态的虚拟机进行操作

图标会变成 ■。可以对处于 Paused 状态的虚拟机执行 Resume 操作,从而使其恢复原有的运行状态。

在 virt-manager 中对虚拟机进行 Shut Down 操作有 5 种模式:Reboot、Shut Down、Force Reset、Force Off 和 Save。

对虚拟机执行 Shut Down 操作,由于执行的是正常的关闭操作,所以不会造成数据的损坏。有 acpi、agent、initctl、signal、paravirt 等多种关闭方法。常用的关闭方法是 acpi 和 agent,其他方法还需要为虚拟机配置额外的参数。在 virt-manager 中,由于没有办法来指定关闭的方法和次序,所以 virt-manager 会尝试采用所有的关闭方法。

提示:使用 virsh edit 命令可以通过编辑虚拟机配置文件中的 on_poweroff 选项,来控制关闭虚拟机的方法。

agent 是指虚拟机中安装的 QEMU Guest Agent,那么 acpi 指的是什么呢?它是高级配置和电源管理接口(Advanced Configuration and Power Management Interface)的缩写。现在的物理计算机或虚拟机通常都有 ACPI 电源按钮设备,如图 4-8 所示。如果快速按下物理计算机的电源按钮,就会向操作系统发出 APCI 关机信号,操作系统接收到此信号就会进行正常关机操作。对于虚拟机来讲,对它执行 Shut Down 操作也是发出 ACPI 关机信号,虚拟机操作系统也会进行正常关机,不会造成数据的损坏。

提示:如果长按物理计算机的电源按钮,则执行的是直接断电操作。

对于 Windows 操作系统来讲,在未登录的情况下,如果收到了 ACPI 关机信号,是否进行关闭还取决于操作系统的配置。这可以通过本地策略中的安全选项进行查看和修改,如图 4-9 所示。

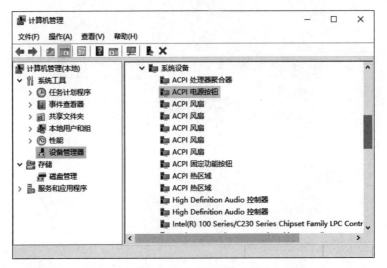

图 4-8　Windows 10 中的 ACPI 电源按钮设备

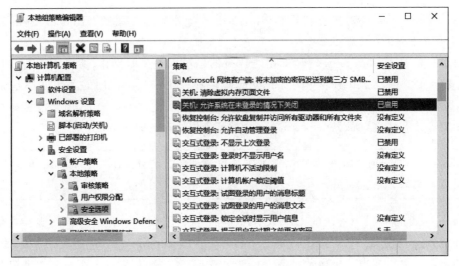

图 4-9　Windows 10 本地策略中的安全选项

此选项对应的注册表键值是:

HKLM \ Software \ Microsoft \ Windows \ CurrentVersion \ Policies \ System \ Shutdownwithoutlogon。

如果 Shutdownwithoutlogon 的值为 0,则表示禁用;如果其值是 1,则表示启用。

对于虚拟机执行 Reboot 操作,相当于先执行 Shut Down 操作,然后执行 Start 操作。

如果无法通过 Shut Down 操作对虚拟机进行正常关闭,则需要使用强制方法。这种方法有可能会造成数据的损坏,所以应尽量避免使用。有两种强制方法:

(1) Force Off:强制关闭虚拟机,相当于直接拔掉物理计算机的电源线。

(2) Force Reset:强制关闭虚拟机并重新启动,相当于按下物理计算机的重置按钮。

注意:对有些虚拟机参数配置进行更改后,需要将虚拟机关闭才会生效。Force Reset 将重置虚拟机,这并不会对任何挂起的虚拟机参数配置进行更改。

Save 是虚拟机特有的关闭方法。首先会保存虚拟机的当前状态(最主要是内存中的数据),然后关闭虚拟机从而释放 CPU 资源。当再启动虚拟机时,会从原来保存的状态中恢复。

4.2.3 管理虚拟硬件

单击 virt-manager 工具栏中的 图标,或者单击 View 菜单中的 Details 子菜单,就可以打开虚拟硬件详细信息窗口。我们可以在其中查看、添加、修改或删除虚拟硬件。

在前面的章节中,已经讲解过了对 CPU、内存、磁盘和图形等虚拟硬件的管理,下面再讲解其他几个选项。

1. 引导选项

可以在 Boot Options 中设置虚拟机在启动时的行为,包括自动启动、引导设备次序和直接内核引导,如图 4-10 所示。

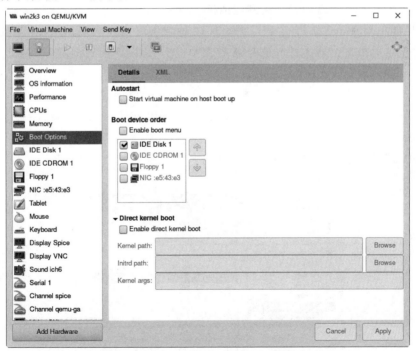

图 4-10　配置虚拟机的引导选项

如果希望虚拟机在宿主机每次启动时也自动启动，则应选中 Start virtual machine on host boot up 复选框。

如果希望虚拟机每次启动时都询问启动的顺序，则应选中 Enable boot menu 复选框。这样此虚拟机启动时就会出现启动菜单，如图 4-11 所示，我们可以通过数字来指定本次引导的设备。

图 4-11　启动虚拟机时指定引导的设备

另外还可以设置默认的引导设备，并使用上下箭头来更改其顺序。

如果要直接从 Linux 内核引导，则需要先展开 Direct Kernel boot，选中 Enable direct Kernel boot，然后填写要使用的 Kernel path、Initrd path 和 Kernel 参数。

2．将宿主机上的 USB 设备连接到虚拟机

虚拟机使用 USB 设备的方法有 2 种。第 1 种方法就是先将 USB 设备连接到宿主机上，然后分配给虚拟机。第 2 种方法是通过软件将 USB 设备重定向给虚拟机。下面以 U 盘为例进行说明。

首先，将 U 盘插入宿主机的 USB 接口，并在宿主机上确认其工作正常，然后给虚拟机添加新的硬件，在 Add New Virtual Hardware 对话框中，选择 USB Host Device，从列表中选择要添加的 USB 设备，最后单击 Finish 按钮关闭对话框，如图 4-12 所示。这时，会在虚拟机配置中看到这个新添加的 USB 设备，如图 4-13 所示。

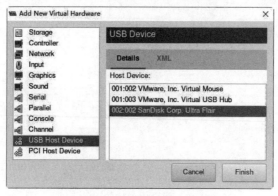

图 4-12　添加 USB 设备

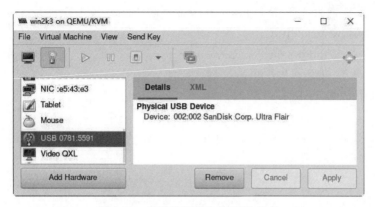

图 4-13　虚拟机配置中的 USB 设备

现在，在虚拟机操作系统中就可以发现并使用这个 U 盘了，如图 4-14 所示。

图 4-14　在虚拟机操作系统中可以发现新增加的 U 盘

3．将 USB 设备重定向给虚拟机

让虚拟机使用 USB 设备的第 2 种方法更灵活一些，它可以将运行 virt-viewer 软件的主机的 USB 设备重定向给虚拟机。这特别适合于宿主机是在远程机房而不在身边的场景。

如果运行 virt-viewer 软件的主机是 Windows 操作系统，则需要在其上安装一个专门为 Spice USB 重定向开发的 Windows 筛选器驱动程序。其下载网址为 https://www.spice-space.org/download.html。

我们可以根据 Windows 操作系统的版本来选择 32 位或 64 位版的驱动程序。下面以 64 位版 Windows 10 为例来做一个简单的实验。

首先，下载并安装 UsbDk_1.0.22_x64.msi，然后将一个 U 盘插入这台计算机的 USB 接口中。

使用 virt-viewer 连接到虚拟机。在网址栏中输入虚拟机的 SPICE 地址，然后单击 Connect 按钮，如图 4-15 所示。

图 4-15　在 Remote viewer 中输入虚拟机的地址

连接成功后，单击"文件"菜单中的 USB device selection，如图 4-16 所示。

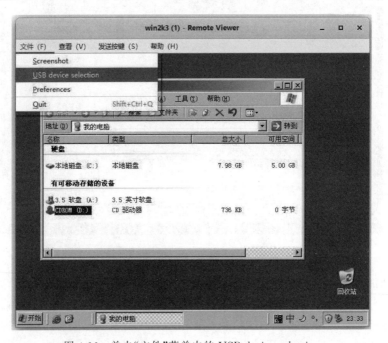

图 4-16　单击"文件"菜单中的 USB device selection

在新的窗口中选择希望重定向到虚拟机的 USB 设备,然后单击 Close 按钮,如图 4-17 所示。

图 4-17　为重新定向选择 USB 设备

这时,在虚拟机操作系统中可以发现新增加的 U 盘了,如图 4-18 所示。

图 4-18　在虚拟机操作系统中可以发现新增加的 U 盘

4.3 使用 virsh 管理虚拟机

virsh 是 Virtual Shell 的缩写,它采用命令行用户界面管理虚拟化平台。我们可以使用 virsh 来创建、列出、编辑、启动、重新启动、停止、挂起、恢复、关闭和删除虚拟机。除了 KVM 之外,它还支持 LXC、Xen、QEMU、OpenVZ、VirtualBox 和 VMware ESX/ESXi。

本章先讲解一些 virsh 最常用的功能。

4.3.1 获得帮助

除了可以使用 man virsh 查看帮助手册之外,我们还可以通过它的 help 选项来查看可用命令列表及简要说明,命令如下:

```
# virsh -- help

virsh [options]... [<command_string>]
virsh [options]... <command> [args...]

  options:
    -c | --connect = URI        hypervisor connection URI
    -d | --debug = NUM          debug level [0-4]
    -e | --escape <char>        set escape sequence for console
    -h | --help                 this help
    -k | --keepalive-interval = NUM
                                keepalive interval in seconds, 0 for disable
    -K | --keepalive-count = NUM
                                number of possible missed keepalive messages
    -l | --log = FILE           output logging to file
    -q | --quiet                quiet mode
    -r | --readonly             connect readonly
    -t | --timing               print timing information
    -v                          short version
    -V                          long version
    --version[=TYPE]  version, TYPE is short or long (default short) commands (non
interactive mode):

Domain Management (help keyword 'domain')
    attach-device         attach device from an XML file
    attach-disk           attach disk device
    attach-interface      attach network interface
    autostart             autostart a domain
    blkdeviotune          Set or query a block device I/O tuning parameters.
    blkiotune             Get or set blkio parameters
    ...
```

我们不必记住所有内容,只需阅读要运行并使用的子命令的说明。这些子命令分为以下几类:

(1) 域管理(Domain Management)。
(2) 域监控(Domain Monitoring)。
(3) 宿主机和管理程序(Host and Hypervisor)。
(4) 接口(Interface)。
(5) 网络过滤器(Network Filter)。
(6) 网络(Networking)。
(7) 节点设备(Node Device)。
(8) 密语(Secret)。
(9) 快照(Snapshot)。
(10) 储存池(Storage Pool)。
(11) 储存卷(Storage Volume)。
(12) virsh 自身的子命令。

提示:在 KVM 中,域(Domain)与虚拟机(Virtual Machine,VM)、快照(Snapshot)和检查点(Checkpoint)是同义词,可以进行互换。

每类子命令都包含与执行特定任务相关的命令。我们可以单独查看某类子命令的帮助,例如查看与虚拟机(域)管理相关子命令的帮助,命令如下:

```
# virsh help domain
Domain Management (help keyword 'domain'):
    attach-device            attach device from an XML file
    attach-disk              attach disk device
    attach-interface         attach network interface
    autostart                autostart a domain
    ...
```

我们还可以进一步显示特定子命令的帮助。例如,查看 list 子命令的帮助,命令如下:

```
# virsh help list
NAME
    list - list domains

SYNOPSIS
    list [--inactive] [--all] [--transient] [--persistent] [--with-snapshot] [--without-snapshot] [--state-running] [--state-paused] [--state-shutoff] [--state-other] [--autostart] [--no-autostart] [--with-managed-save] [--without-managed-save] [--uuid] [--name] [--table] [--managed-save] [--title]
```

```
    DESCRIPTION
        Returns list of domains.

    OPTIONS
        --inactive          list inactive domains
        --all               list inactive & active domains
        --transient         list transient domains
...
```

4.3.2 常用的子命令

1. 列出虚拟机

要获得处于运行或挂起(suspend)模式的虚拟机的列表,命令如下:

```
#virsh list
 Id    Name                           State
----------------------------------------------------
 1     CentOS6.10                     running
```

可以使用--inactive 选项只显示不活动的虚拟机。

如果想查看所有状态的虚拟机,则应执行的命令如下:

```
#virsh list --all
 Id    Name                           State
----------------------------------------------------
 1     CentOS6.10                     running
 -     CentOS6.10vm2                  shut off
 -     CentOS6.10vm3                  shut off
 -     CentOS8                        shut off
 -     demo-guest1                    shut off
 -     win2k19                        shut off
 -     win2k3                         shut off
 -     win2k3-2                       shut off
 -     win2k8                         shut off
```

我们可以从输出信息中看到:有 1 台处于 running 状态的虚拟机,libvirt 给它分配的 ID 号是 1。其他虚拟机的状态是 shut off,而且没有 ID 号。

2. 启动虚拟机

要启动虚拟机,例如启动名称为 win2k3 的虚拟机,示例命令如下:

```
#virsh start win2k3
Domain win2k3 started
```

> 提示：virsh 具有自动完成功能。我们可以使用键盘上的 Tab 键来完成子命令、选项和参数的自动补全，例如上述命令中的 start 子命令和 win2k3 参数。

验证虚拟机是否正在运行，命令如下：

```
# virsh list
Id    Name                State
----------------------------------------------------
1     centos6.10          running
2     win2k3              running
```

3. 关闭虚拟机

使用 shutdown 子命令可以正常地关闭虚拟机。在使用时，我们既可以通过名称也可以通过 ID 来指定虚拟机，命令如下：

```
# virsh list
Id    Name                State
----------------------------------------------------
2     win2k3              running
3     centos6.10          running

# virsh shutdown 3
Domain 3 is being shutdown

# virsh list
Id    Name                State
----------------------------------------------------
2     win2k3              running
```

由于 shutdown 子命令需要与虚拟机的操作系统一起协调工作，所以不能保证它会成功地关闭虚拟机，而且关闭所需要时间的长短，也取决于虚拟机中要关闭的服务和进程的多少。

默认情况下，Hypervisor 会尝试选择合适的关闭方法。我们也可以通过 --mode 参数指定关闭方法的列表，可以是 acpi、agent、initctl、signal 和 paravirt 中的一种或多种。

如果 shutdown 子命令无法关闭虚拟机，则我们只能通过 destroy 子命令来强制关闭虚拟机了，这与把电源线直接从物理机上拔出很类似，命令如下：

```
# virsh shutdown win2k3
Domain win2k3 is being shutdown
```

如果等待一段时间后，虚拟机还在运行，则说明 shutdown 子命令执行失败，命令如下：

```
# virsh list
 Id    Name                    State
----------------------------------------------------
 2     win2k3                  running

# virsh destroy win2k3
Domain win2k3 destroyed
```

提示：每台运行着的虚拟机都是宿主机上的一个进程(/usr/libexec/qemu-kvm)，但是不推荐使用 kill 命令结束进程的方法来关闭虚拟机。

4. 重新启动虚拟机

使用 reboot 子命令可以重新启动虚拟机，就像从控制台中运行重新启动命令一样。这个子命令当然也需要与虚拟机的操作系统协同工作，命令如下：

```
# virsh reboot centos6.10
Domain centos6.10 is being rebooted
```

如果 reboot 子命令无法重新启动虚拟机，则我们只能通过 reset 子命令来强制重新启动虚拟机了，这与按下物理机上的重置按钮类似。示例命令如下：

```
# virsh reset win2k3
Domain win2k3 was reset
```

注意：destroy 和 reset 子命令有可能会造成虚拟机数据的损坏，需谨慎使用。

5. 挂起(暂停)与恢复虚拟机

使用 suspend 子命令可以挂起(暂停)正在运行的虚拟机，而使用 resume 子命令又可以将它们恢复运行。命令如下：

```
# virsh list
 Id    Name                    State
----------------------------------------------------
 1     centos6.10              running

# virsh suspend centos6.10
Domain centos6.10 suspended

# virsh list
 Id    Name                    State
----------------------------------------------------
 1     centos6.10              paused
```

```
# virsh resume centos6.10
Domain centos6.10 resumed

# virsh list
 Id    Name                         State
----------------------------------------------------
 1     centos6.10                   running
```

6. 保存与还原虚拟机

可以将正在运行的虚拟机的内存数据保存到状态文件(State File)中，这类似于在虚拟机操作系统中进行休眠操作。保存虚拟机之后，该虚拟机将不再在宿主机上运行，因此会释放原来分配给该虚拟机的资源。保存虚拟机的命令如下：

```
# virsh save centos6.10 centos6.10-save
Domain centos6.10 saved to centos6.10-save

# ls -lh centos6.10-save
-rw-------. 1 root root 189M Oct 25 18:29 centos6.10-save

# file centos6.10-save
centos6.10-save: Libvirt QEMU Suspend Image, version 2, XML length 5562, running

# virsh list --all
 Id    Name                         State
----------------------------------------------------
 2     win2k3                       running
 -     centos6.10                   shut off
 -     centos6.10vm2                shut off
 -     centos6.10vm3                shut off
 -     centos8                      shut off
 -     demo-guest1                  shut off
 -     win2k19                      shut off
 -     win2k3-2                     shut off
 -     win2k8                       shut off
```

保存虚拟机需要一些时间，这取决于虚拟机内存数据的多少。

当需要虚拟机的时候，可以进行还原操作。例如，从状态文件 centos6.10-save 还原虚拟机的命令如下：

```
# virsh restore centos6.10-save
Domain restored from centos6.10-save

# virsh list
```

```
 Id    Name                State
----------------------------------------------
 2     win2k3              running
 3     centos6.10          running
```

7. 查看编辑虚拟机配置文件

默认情况下，libvirtd 使用保存在/etc/libvirt/qemu/目录中的虚拟机配置文件，命令如下：

```
# ls -l /etc/libvirt/qemu/
total 72
-rw-------. 1 root root 4749 Aug  4 23:36 centos6.10vm2.xml
-rw-------. 1 root root 4651 Aug  3 22:29 centos6.10vm3.xml
-rw-------. 1 root root 4839 Sep  2 20:25 centos6.10.xml
-rw-------. 1 root root 5896 Aug 24 20:32 centos8.xml
-rw-------. 1 root root 5061 Aug 20 22:16 demo-guest1.xml
drwx------. 3 root root   42 Jul  6 20:52 networks
-rw-------. 1 root root 6006 Aug 23 11:11 win2k19.xml
-rw-------. 1 root root 4740 Aug 15 22:34 win2k3-2.xml
-rw-------. 1 root root 5299 Sep  2 22:41 win2k3.xml
-rw-------. 1 root root 4759 Aug 14 22:13 win2k8.xml
```

一个虚拟机对应于一个 XML 格式的配置文件。除了可以使用 cat 等命令来查看配置文件之外，还可以使用 virsh 的 dumpxml 子命令来查看虚拟机的当前配置，命令如下：

```
# virsh dumpxml centos6.10
<domain type='kvm'>
  <name>centos6.10</name>
  <uuid>edcba2b1-afa9-4de5-8de4-b42cc1bc66c9</uuid>
  <metadata>
    <libosinfo:libosinfo xmlns:libosinfo="http://libosinfo.org/xmlns/libvirt/domain/1.0">
      <libosinfo:os id="http://centos.org/centos/6.10"/>
    </libosinfo:libosinfo>
  </metadata>
  <memory unit='KiB'>1048576</memory>
  <currentMemory unit='KiB'>1048576</currentMemory>
  <vcpu placement='static'>1</vcpu>
  <os>
    <type arch='x86_64' machine='pc-i440fx-rhel7.6.0'>hvm</type>
    <boot dev='hd'/>
    <bootmenu enable='yes'/>
  </os>
  ...
```

配置文件中元素和属性的详细介绍可参阅 libvirt 的官方文档：https://libvirt.org/formatdomain.html。

我们可以通过编辑配置文件来修改虚拟机的设置，这种修改需要在虚拟机下次启动时才会生效。不建议通过 vi 等文本编辑软件直接修改配置文件，而是要使用 virsh 的 edit 子命令。edit 子命令具有一些错误检查的功能，命令如下：

```
# virsh edit CentOS6.10
```

edit 子命令将使用由 $EDITOR 变量设置的编辑器打开文件，默认为 vi 编辑器。

提示：edit 子命令会自动将配置文件复制为一个新的临时文件，然后调用文本编辑器打开这个文件。当保存修改时，会先进行格式检查，没有问题之后才再覆盖原有的配置文件。

8. 创建新的虚拟机

除了可以使用 virt-install、virt-manager 和 Cockpit 创建新的虚拟机之外，还可以使用 virsh 的 create 或 define 子命令来创建新的虚拟机。

create 子命令可以创建新的临时性虚拟机，如果同时指定了 --paused 选项，则新虚拟机的状态为暂停状态，否则将处于运行状态。由于 create 子命令创建的是临时虚拟机，所以不会在 /etc/libvirt/qemu/ 目录中创建新虚拟机的配置文件。一旦关闭临时虚拟机，libvirt 就不知道它的存在了。

与 create 命令正好相反，define 子命令会在 /etc/libvirt/qemu/ 目录中创建新虚拟机的配置文件，所以是永久性虚拟机，但是新虚拟机处于关闭状态，还需要使用 start 子命令来启动它。

create 和 define 子命令都需要使用一个 XML 格式的配置文件。我们既可以根据 libvirt 的 XML 格式要求（https://www.libvirt.org/format.html）来全新创建，也可以参考现有的虚拟机的配置文件来创建。最简单的方法是使用 dumpxml 从现有的虚拟机中先复制出一个基础的配置，然后进行修改。下面，我们来做一个实验：

```
1 # virsh dumpxml centos6.10 > newvm1.xml

2 # virsh domblklist centos6.10
  Target  Source
  ------------------------------------------------
  vda     /vm/centos6.10.qcow2
  hda     -

3 # cp /vm/centos6.10.qcow2 /vm/newvm1.qcow2

4 # vi newvm1.xml
```

为了避免两台虚拟机使用相同的虚拟磁盘文件，所以通过第 2 行命令获得原虚拟机的磁盘文件的信息，第 3 行命令复制出一个新的磁盘文件供新的虚拟机使用。

在编辑 XML 文件时，要注意新虚拟机的名称要保持唯一，将 UUID 删除，同时使用新的虚拟磁盘文件，代码如下：

```
< domain type = 'kvm'>
  < name > newvm1 </name >
  <uuid> ba0f2c14 – 24ee – 44c2 – a0db – 9ae55914b461 </uuid>
  ...
    < disk type = 'file' device = 'disk'>
      < driver name = 'qemu' type = 'qcow2'/>
      < source file = '/vm/newvm1.qcow2'/>
      < target dev = 'vda' bus = 'virtio'/>
      < address type = 'pci' domain = '0x0000' bus = '0x00' slot = '0x07' function = '0x0'/>
    </disk>
  ...
```

提示：可以先通过 virt-xml-validate 命令对配置文件进行检查。

有了这个新的 XML 文件，下面就可以创建新的虚拟机了。我们先尝试使用 create 子命令来创建新的虚拟机，命令如下：

```
5 # virsh create newvm1.xml
  Domain newvm1 created from newvm1.xml

6 # virsh list
    Id    Name                    State
    ----------------------------------------
    7     newvm1                  running

7 # ls /etc/libvirt/qemu/
  centos6.10.xml networks win2k19.xml win2k3.xml

8 # virsh destroy newvm1
  Domain newvm1 destroyed

9 # virsh list -- all
    Id    Name                    State
    ----------------------------------------
    -     centos6.10              shut off
    -     win2k19                 shut off
    -     win2k3                  shut off
```

从第 7 行命令的输出可以看出：create 子命令并不会创建新虚拟机的配置文件，所以，如果将虚拟机关闭,libvirt 就不会知道有过这个虚拟机,这可以从第 9 行命令的输出中得到验证。

接下来,我们使用 define 子命令来创建新的虚拟机,命令如下:

```
10 # virsh define newvm1.xml
   Domain newvm1 defined from newvm1.xml

11 # virsh list -- all
    Id     Name                    State
   ----------------------------------------------------
    -      centos6.10              shut off
    -      newvm1                  shut off
    -      win2k19                 shut off
    -      win2k3                  shut off

12 # ls -lt /etc/libvirt/qemu/
   total 32
   -rw-------. 1 root root 4610 Nov 6 16:48 newvm1.xml
   -rw-------. 1 root root 5364 Nov 6 13:36 win2k19.xml
   -rw-------. 1 root root 4622 Nov 5 15:28 CentOS6.10.xml
   -rw-------. 1 root root 4663 Nov 4 17:39 win2k3.xml
   drwx------. 3 root root 42 Jul 30 16:40 networks

13 # virsh start newvm1
   Domain newvm1 started
```

从第 11 行和第 12 行命令的输出可以看出：define 子命令创建的新虚拟机处于关闭状态,但是在 /etc/libvirt/qemu/ 目录中会有相关配置文件。

9. 删除虚拟机

如果不再需要某个虚拟机,我们则可以先将其关闭,然后使用 undefine 子命令将配置文件删除。如果虚拟机正在运行,则 undefine 子命令不会停止它。默认情况下,undefine 子命令不会删除虚拟机的磁盘文件,命令如下:

```
1 # virsh destroy newvm1
   Domain newvm1 destroyed

2 # virsh undefine newvm1
   Domain newvm1 has been undefined

3 # virsh list -- all | grep newvm1

4 # rm /vm/newvm1.qcow2
   rm: remove regular file '/vm/newvm2.qcow2'? y
```

> 提示：如果虚拟机的磁盘文件是通过 libvirt 的存储池、存储卷进行管理的，则还可以使用 undefine 子命令中的 --remove-all-storage、--storage 选项来同时清除它们。

4.4 使用 Cockpit 管理虚拟机

Cockpit 为我们提供了非常友好的 Web 前端界面。它简单易用，无须阅读帮助文件就可以发现如何通过 Cockpit 来执行管理任务，如果有 virt-manager 和 virsh 的使用经验，就可以更轻松地管理虚拟化平台了。

Cockpit 直接与系统交互，没有自己的私有的数据，所以界面上始终反映服务器当前的实际状态，例如 Cockpit 向虚拟机提供的数据来源于 QEMU/libvirt，通过 virsh 工具或 libvirt D-Bus API 进行管理。

目前，我们可以通过 cockpit-machines 插件来完成以下操作：

（1）查看可用虚拟机的列表。
（2）查看虚拟机的详细信息。
（3）编辑虚拟机参数，例如内存、磁盘和网络接口等。
（4）创建新虚拟机。
（5）导入虚拟机。
（6）运行虚拟机。
（7）删除虚拟机。
（8）查看存储池。
（9）创建存储池。
（10）编辑存储池。
（11）查看默认虚拟网络。
（12）创建新的虚拟网络。
（13）编辑虚拟网络。
（14）删除虚拟网络。

4.5 本章小结

本章讲解了虚拟化管理的引擎 libvirt，如何通过 virt-manager、virsh 和 Cockpit 管理虚拟机，包括创建、暂停、恢复、停止及删除等生命周期的操作管理。

第 5 章将讲解管理虚拟网络。

第 5 章 管理虚拟网络

本章将系统地讲解虚拟网络的日常管理，这需要具备一定的 Linux 网络管理的知识，包括物理网络接口、虚拟网络接口、网桥及路由等。

本章要点：
- TUN/TAP 设备工作原理与管理。
- 网桥工作原理与管理。
- 理解不同网络类型的原理。
- 掌握 NAT、桥接、隔离、路由、开放等网络类型的配置。
- 掌握 VLAN 的原理与配置。
- 掌握网络过滤器的原理与配置。

5.1 查看默认网络环境

KVM 支持多种网络类型，首先我们查看一下默认的网络环境。

在 RHEL/CentOS 8 等 Linux 发行版本中安装虚拟化组件的时候，通常会自动创建一个默认的虚拟网络配置，这包括一个名为 virbr0 的网桥、virbr0-nic 的虚拟网络接口、iptables 的 NAT 配置及 DNSMASQ 的配置等。下面我们就通过实验来查看默认的网络环境，从而理解 libvirt 虚拟网络的原理。

5.1.1 查看宿主机的网络环境

在没有启动虚拟机的情况下，宿主机上默认的网络环境如图 5-1 所示。

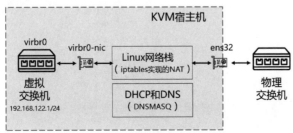

图 5-1　宿主机上默认的网络环境

查看宿主机网络环境的命令如下：

```
1 # ip address
  1: lo: <LOOPBACK,UP,LOWER_UP> mtu 65536 qdisc noqueue state UNKNOWN group default qlen 1000
      link/loopback 00:00:00:00:00:00 brd 00:00:00:00:00:00
      inet 127.0.0.1/8 scope host lo
         valid_lft forever preferred_lft forever
      inet6 ::1/128 scope host
         valid_lft forever preferred_lft forever
  2: ens32: <BROADCAST,MULTICAST,UP,LOWER_UP> mtu 1500 qdisc fq_codel state UP group default qlen 1000
      link/ether 00:0c:29:f7:6b:c8 brd ff:ff:ff:ff:ff:ff
      inet 192.168.114.231/24 brd 192.168.114.255 scope global noprefixroute ens32
         valid_lft forever preferred_lft forever
      inet6 fe80::20c:29ff:fef7:6bc8/64 scope link
         valid_lft forever preferred_lft forever
  3: virbr0: <NO-CARRIER,BROADCAST,MULTICAST,UP> mtu 1500 qdisc noqueue state DOWN group default qlen 1000
      link/ether 52:54:00:e0:41:ac brd ff:ff:ff:ff:ff:ff
      inet 192.168.122.1/24 brd 192.168.122.255 scope global virbr0
         valid_lft forever preferred_lft forever
  4: virbr0-nic: <BROADCAST,MULTICAST> mtu 1500 qdisc fq_codel master virbr0 state DOWN group default qlen 1000
      link/ether 52:54:00:e0:41:ac brd ff:ff:ff:ff:ff:ff

2 # ip link
  1: lo: <LOOPBACK,UP,LOWER_UP> mtu 65536 qdisc noqueue state UNKNOWN mode DEFAULT group default qlen 1000
      link/loopback 00:00:00:00:00:00 brd 00:00:00:00:00:00
  2: ens32: <BROADCAST,MULTICAST,UP,LOWER_UP> mtu 1500 qdisc fq_codel state UP mode DEFAULT group default qlen 1000
      link/ether 00:0c:29:f7:6b:c8 brd ff:ff:ff:ff:ff:ff
  3: virbr0: <NO-CARRIER,BROADCAST,MULTICAST,UP> mtu 1500 qdisc noqueue state DOWN mode DEFAULT group default qlen 1000
      link/ether 52:54:00:e0:41:ac brd ff:ff:ff:ff:ff:ff
  4: virbr0-nic: <BROADCAST,MULTICAST> mtu 1500 qdisc fq_codel master virbr0 state DOWN mode DEFAULT group default qlen 1000
      link/ether 52:54:00:e0:41:ac brd ff:ff:ff:ff:ff:ff

3 # nmcli connection
  NAME    UUID                                    TYPE      DEVICE
  ens32   152beb06-47c5-c5e8-95a9-385590654382    ethernet  ens32
  virbr0  3068816f-b57c-4620-aea2-0e83642cdd87    bridge    virbr0

4 # nmcli device
  DEVICE       TYPE       STATE       CONNECTION
  ens32        ethernet   connected   ens32
  virbr0       bridge     connected   virbr0
  lo           loopback   unmanaged   --
  virbr0-nic   tun        unmanaged   --
```

libvirt 网络的最重要的组件是虚拟网络交换机，默认为由 Linux 的网桥实现。libvirt 默认会创建一个名为 virbr0 的网桥，virbr0 是 Virtual Bridge 0 的缩写。与物理交换机类似，虚拟网络交换机也是从其接收的数据包(帧)中获得 MAC 地址，并存储在 MAC 表中。物理交换机的端口数量有限，而虚拟交换机的端口数量则没有限制。

在 Linux 系统中，可以向网桥分配 IP 地址。从第 1 行命令的输出中可以看到，libvirt 向 virbr0 分配的 IP 地址是 192.168.122.1/24。libvirt 根据自己的配置文件来分配这个 IP 地址，所以不会在/etc/sysconfig/network-scripts/目录中看到名为 ifcfg-virbr0 的配置文件。

由于 Linux 的网桥会将其上第 1 个接口设备的 MAC 地址当作它的 MAC 地址，所以 libvirt 会创建一个名为 virbr0-nic 的 TAP 设备。从第 1 行、第 2 行命令的输出中可以看出 virbr0 与 virbr0-nic 的 MAC 地址都是 52:54:00:e0:41:ac。

连接网桥的接口通常被称为 slave 接口。它们既可以是物理网卡，也可以是虚拟网络接口设备(例如：TAP 类型的虚拟设备)，所以通过网桥可以连通物理与虚拟网络设备。从相互关系上来讲，这些 slave 接口的 master 就是虚拟机交换机。从第 1 行、第 2 行命令的输出中可以看到 virbr0-nic 的属性字符串有 master virbr0 字样，这说明它是 virbr0 网桥的 slave 接口。

5.1.2 查看 libvirt 的网络环境

在 libvirtd 守护程序启动时，会根据配置文件创建一个名为 default 的虚拟网络。除了在宿主机上创建 virbr0、virbr0-nic 之外，还会通过配置 IP 转发、iptables 的 NAT 表从而在 Linux 协议栈中实现 NAT 功能。

libvirt 使用的是 IP 伪装(IP masquerading)，而不是源 NAT(Source-NAT，SNAT)或目标 NAT(Destination-NAT，DNAT)。IP 伪装使虚拟机可以使用宿主机的 IP 地址与外部网络进行通信。默认情况下，当虚拟网络交换机以 NAT 模式运行时，虚拟机可以访问位于宿主物理计算机外部的资源，但是位于宿主物理计算机外部的计算机无法与内部的虚拟机进行通信，也就是说：仅允许由内到外的访问，而不允许从外到内的访问。

首先，我们通过 virt-manager 这种比较直观的方式来查看虚拟网络 default：

(1) 在 virt-manager 的 Edit 菜单上，选择 Connection Details。
(2) 在打开的 Connection Details 菜单中单击 Virtual Networks 选项卡，如图 5-2 所示。
(3) 窗口的左侧列出了所有可用的虚拟网络。我们可以查看和修改虚拟网络的配置。

下面通过命令行工具查看网络环境，命令如下：

```
1 # cat /proc/sys/net/ipv4/ip_forward
  1

2 # iptables -t nat -L -n
  Chain PREROUTING (policy ACCEPT)
  target     prot opt source               destination
```

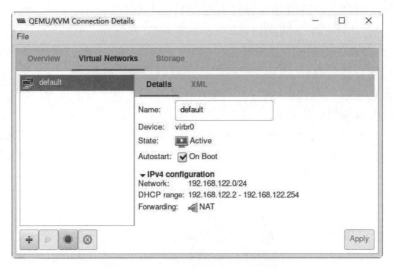

图 5-2　virt-manager 中的虚拟网络配置

```
Chain INPUT (policy ACCEPT)
target     prot opt source              destination

Chain POSTROUTING (policy ACCEPT)
target     prot opt source              destination
RETURN     all  --  192.168.122.0/24    224.0.0.0/24
RETURN     all  --  192.168.122.0/24    255.255.255.255
MASQUERADE tcp  --  192.168.122.0/24    !192.168.122.0/24   masq ports: 1024-65535
MASQUERADE udp  --  192.168.122.0/24    !192.168.122.0/24   masq ports: 1024-65535
MASQUERADE all  --  192.168.122.0/24    !192.168.122.0/24

Chain OUTPUT (policy ACCEPT)
target     prot opt source              destination
```

注意：不建议在虚拟交换机运行时编辑这些防火墙规则，因为有可能导致交换机通信故障。

为了简化虚拟网络中的管理，libvirt 还使用了 DNSMASQ 组件为 default 网络中的虚拟机提供 DNS 和 DHCP 功能，命令如下：

```
3 # ps aux | grep dnsmasq
  dnsmasq     1561  0.0  0.0  71888  2428 ?        S    08:21   0:00
/usr/sbin/dnsmasq --conf-file=/var/lib/libvirt/dnsmasq/default.conf
--leasefile-ro --dhcp-script=/usr/libexec/libvirt_leaseshelper
  root        1562  0.0  0.0  71860  1828 ?        S    08:21   0:00
/usr/sbin/dnsmasq --conf-file=/var/lib/libvirt/dnsmasq/default.conf
```

```
    --leasefile-ro --dhcp-script=/usr/libexec/libvirt_leaseshelper
    root      3041  0.0  0.0  12108  1108 pts/2     S+   09:24    0:00 grep
    --color=auto dnsmasq

4 # cat /var/lib/libvirt/dnsmasq/default.conf
    ##WARNING: THIS IS AN AUTO-GENERATED FILE. CHANGES TO IT ARE LIKELY TO BE
    ##OVERWRITTEN AND LOST. Changes to this configuration should be made using:
    ## virsh net-edit default
    ##or other application using the libvirt API.
    ##
    ##dnsmasq conf file created by libvirt
    strict-order
    pid-file=/var/run/libvirt/network/default.pid
    except-interface=lo
    bind-dynamic
    interface=virbr0
    dhcp-range=192.168.122.2,192.168.122.254
    dhcp-no-override
    dhcp-authoritative
    dhcp-lease-max=253
    dhcp-hostsfile=/var/lib/libvirt/dnsmasq/default.hostsfile
    addn-hosts=/var/lib/libvirt/dnsmasq/default.addnhosts
```

从第 3 行命令的输出中可以看出 DNSMASQ 所使用的配置文件也是由 libvirt 提供的，它的路径名为 /var/lib/libvirt/dnsmasq/default.conf。

从第 4 行命令的输出中可以看出向虚拟机分配的 IP 地址的范围是从 192.168.122.2 到 192.168.122.254。

接下来，执行如下命令：

```
5 # virsh net-list
    Name         State      Autostart    Persistent
    ----------------------------------------------------------
    default      active     yes          yes

6 # virsh net-dumpxml default
    <network connections='2'>
      <name>default</name>
      <uuid>6c729bec-ce6c-4ca3-b05c-1fdb99be4fdc</uuid>
      <forward mode='nat'>
        <nat>
          <port start='1024' end='65535'/>
        </nat>
      </forward>
      <bridge name='virbr0' stp='on' delay='0'/>
```

```
        < mac address = '52:54:00:e0:41:ac'/>
        < ip address = '192.168.122.1' netmask = '255.255.255.0'>
          < dhcp >
            < range start = '192.168.122.2' end = '192.168.122.254'/>
          </dhcp >
        </ ip >
      </network >

7  # ls /etc/libvirt/qemu/networks/
   autostart default.xml

8  # cat /etc/libvirt/qemu/networks/default.xml
   <!--
   WARNING: THIS IS AN AUTO-GENERATED FILE. CHANGES TO IT ARE LIKELY TO BE
   OVERWRITTEN AND LOST. Changes to this xml configuration should be made using:
       virsh net-edit default
   or other application using the libvirt API.
   -->

   < network >
     < name > default </name >
     < uuid > 6c729bec-ce6c-4ca3-b05c-1fdb99be4fdc </uuid >
     < forward mode = 'nat'/>
     < bridge name = 'virbr0' stp = 'on' delay = '0'/>
     < mac address = '52:54:00:e0:41:ac'/>
     < ip address = '192.168.122.1' netmask = '255.255.255.0'>
       < dhcp >
         < range start = '192.168.122.2' end = '192.168.122.254'/>
       </dhcp >
     </ ip >
   </network >
```

从第 5 行命令的输出可以看出，这个名为 default 的网络是随着 libvirtd 的启动而自动启动的，当前的状态是已激活。

第 6 行命令输出了 default 网络的详细定义：

(1) < name > default </name > 指定了虚拟网络的名称。

(2) < uuid > 6c729bec-ce6c-4ca3-b05c-1fdb99be4fdc </uuid > 指定了虚拟网络的全局唯一标识符。

(3) < forward mode＝'nat'> 指定了虚拟网络将连接到物理网络，mode 属性确定了转发方法，目前允许的方法有 nat、route、open、bridge、private、vepa、passthrough 和 hostdev。如果没有配置 forward 属性，则该网络与任何其他网络都是隔离的，也就是 isolated 模式。nat 会在连接到该网络的虚拟机与物理网络之间进行网络地址转换。

(4) < port start＝'1024' end＝'65535'/> 设置了用于< nat >的端口范围。

（5）< bridge name＝'virbr0' stp＝'on' delay＝'0'/>指定了 libvirt 在宿主机上创建网桥设备的信息。name 属性定义了网桥设备的名称，新网桥启用对生成树协议（STP）的支持，默认延迟为 0。建议使用以 virbr 开头的桥名称。虚拟机连接到该桥接设备，就像将真实世界的计算机连接到物理交换机一样。

（6）< mac address＝'52:54:00:e0:41:ac'/>属性定义了一个 MAC（硬件）地址，格式为 6 组 2 位十六进制数字，各组之间用冒号分隔。这个 MAC 地址在创建时即分配给桥接设备。建议让 libvirt 自动生成一个随机 MAC 地址并将其保存在配置中。

（7）< ip address＝'192.168.122.1' netmask＝'255.255.255.0'>为网桥指定 IP 地址。

（8）< dhcp >指定了在虚拟网络上启用 DHCP 服务。

（9）< range start＝'192.168.122.2' end＝'192.168.122.254'/>指定了要提供给 DHCP 客户端的地址池的边界。

提示：libvirt 网络配置的详细介绍可参见 https://libvirt.org/formatnetwork.html。

RHEL/CentOS 8 中的 libvirt 网络配置保存在/etc/libvirt/qemu/networks/目录下的 XML 文件中。强烈不建议直接编辑这些配置文件，而应通过 virsh 的 net-edit 子命令来修改。

5.1.3　查看虚拟机的网络配置

下面，我们再查看一下当启动虚拟机后网络所发生的变化，命令如下：

```
1 # virsh domiflist centos6.10
 Interface   Type     Source    Model    MAC
 ---------------------------------------------------
 -           network  default   virtio   52:54:00:27:5f:c9

2 # virsh start centos6.10

3 # virsh domiflist centos6.10
 Interface   Type     Source    Model    MAC
 ---------------------------------------------------
 vnet0       network  default   virtio   52:54:00:27:5f:c9
```

从第 1 行命令的输出中可以看出：这个名为 centos6.10 的虚拟机未启动之时，Interface 的属性为空，它连接到网络的名称是 default。

当启动此虚拟机时，libvirt 会在宿主机上创建一个新的虚拟网络接口，并将其连接到网桥中的 virbr0 上。我们可以从系统日志（/var/log/messages）中看到类似这样的信息：

```
Kernel: virbr0: port 2(vnet0) entered blocking state
Kernel: virbr0: port 2(vnet0) entered disabled state
Kernel: device vnet0 entered promiscuous mode
```

新的虚拟网络接口的名称是以 vnet 开头的，后面是从 0 开始的序号。如果虚拟机操作系统的网络是自动获得 IP 地址的，则还会在系统日志中看到它从 DNSMASQ 中租用 IP 地址的信息：

```
dnsmasq-dhcp[1561]: DHCPDISCOVER(virbr0) 192.168.122.142 52:54:00:27:5f:c9
dnsmasq-dhcp[1561]: DHCPOFFER(virbr0) 192.168.122.142 52:54:00:27:5f:c9
dnsmasq-dhcp[1561]: DHCPREQUEST(virbr0) 192.168.122.142 52:54:00:27:5f:c9
dnsmasq-dhcp[1561]: DHCPACK(virbr0) 192.168.122.142 52:54:00:27:5f:c9
```

我们还可以使用 ip 和 nmcli 命令查看宿主机上网络的变化情况，命令如下：

```
4 # ip addr
    …
    3: virbr0: <BROADCAST,MULTICAST,UP,LOWER_UP> mtu 1500 qdisc noqueue state UP group default qlen 1000
        link/ether 52:54:00:e0:41:ac brd ff:ff:ff:ff:ff:ff
        inet 192.168.122.1/24 brd 192.168.122.255 scope global virbr0
           valid_lft forever preferred_lft forever
    4: virbr0-nic: <BROADCAST,MULTICAST> mtu 1500 qdisc fq_codel master virbr0 state DOWN group default qlen 1000
        link/ether 52:54:00:e0:41:ac brd ff:ff:ff:ff:ff:ff
    5: vnet0: <BROADCAST,MULTICAST,UP,LOWER_UP> mtu 1500 qdisc fq_codel master virbr0 state UNKNOWN group default qlen 1000
        link/ether fe:54:00:27:5f:c9 brd ff:ff:ff:ff:ff:ff
        inet6 fe80::fc54:ff:fe27:5fc9/64 scope link
           valid_lft forever preferred_lft forever

5 # nmcli connection
NAME     UUID                                  TYPE      DEVICE
ens32    152beb06-47c5-c5e8-95a9-385590654382  ethernet  ens32
virbr0   3068816f-b57c-4620-aea2-0e83642cdd87  bridge    virbr0
vnet0    0c886394-b825-432f-9f79-bfe0a67199db  tun       vnet0
```

从第 4 行命令的输出可以看出：新增加了一个名 vnet0 的接口，它的 master 是 virbr0。第 5 行命令的输出显示了它的类型是 TUN，准确来讲应当是 TAP 设备。

注意：NetworkManager 有些版本（例如 1.22.8-5.el8_2.x86_64）总是将 TUN/TAP 设备显示为 TAP 设备，而 ip 命令（例如 iproute-5.3.0-1.el8.x86_64）显示的类型则是正确的。5.2 节将深入讲解 TUN/TAP 设备。

我们再启动一台新的虚拟机 win2k3，它会在网桥 virbr0 中再新增加一个 slave 设备，新设备名称为 vnet1，命令如下：

```
6 # ip addr
    ...
 5: vnet0: < BROADCAST,MULTICAST,UP,LOWER_UP > mtu 1500 qdisc fq_codel master virbr0 state
UNKNOWN group default qlen 1000
        link/ether fe:54:00:27:5f:c9 brd ff:ff:ff:ff:ff:ff
        inet6 fe80::fc54:ff:fe27:5fc9/64 scope link
           valid_lft forever preferred_lft forever
 6: vnet1: < BROADCAST,MULTICAST,UP,LOWER_UP > mtu 1500 qdisc fq_codel master virbr0 state
UNKNOWN group default qlen 1000
        link/ether fe:54:00:9d:57:9c brd ff:ff:ff:ff:ff:ff
        inet6 fe80::fc54:ff:fe9d:579c/64 scope link
           valid_lft forever preferred_lft forever

7 # nmcli connection
   NAME     UUID                                    TYPE       DEVICE
   ens32    152beb06-47c5-c5e8-95a9-385590654382    ethernet   ens32
   virbr0   3068816f-b57c-4620-aea2-0e83642cdd87    bridge     virbr0
   vnet0    0c886394-b825-432f-9f79-bfe0a67199db    tun        vnet0
   vnet1    3d4e1388-aee6-42df-a283-fcc909939762    tun        vnet1
```

此时宿主机上网络环境如图 5-3 所示。两台虚拟机之间的通信、宿主机与虚拟机之间的通信全部是在虚拟交换机（网桥）virbr0 内部完成的。两台虚拟机对外部网络的访问是先将消息发送给它们的默认网关 virbr0(192.168.122.1)，再通过 iptables 实现的 NAT 功能从 ens32 发送出去。由于采用的 NAT 模式是 MASQUERADE（一种 SNAT），所以默认情况下外部网络的主机不能访问这两台虚拟机，换句话来讲，就是这两台虚拟机对于外部网络来讲是不可见的。

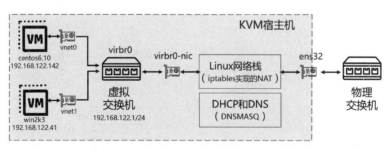

图 5-3 RHEL/CentOS8 中 libvirt 使用的 default 网络的工作原理

5.2 TUN/TAP 设备工作原理与管理

libvirt 的虚拟网络离不开 TAP 类型的虚拟网络接口设备，所以需要讲解一下它的工作原理。首先回顾一下 Linux 物理网卡的工作原理。计算机系统通常会有一个或多个网络设备，例如 eth0、eth1 等。这些网络设备与物理网络适配器相关联，后者负责将数据包放置到线路上，如图 5-4 所示。

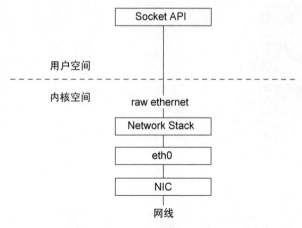

图 5-4　Linux 物理网卡工作原理

eth0 通过物理网卡 NIC 与外部网络相连，该物理网卡收到的数据包会经由 eth0 传递给内核的网络协议栈（Network Stack），然后由协议栈对这些数据包进行进一步的处理。

对于一些错误的数据包，协议栈可以选择丢弃。对于目标不属于本机的数据包，协议栈可以选择转发，而对于目标地址是本机的而且是上层应用所需要数据包，协议栈会通过 Socket API 传递给上层正在等待的应用程序。

虚拟网络接口设备完成的功能与物理网卡类似，有两种常见类型：TUN 和 TAP。在 Linux 内核官方网站上有一个文档（https://www.Kernel.org/doc/Documentation/networking/tuntap.txt）是这样定义的：TUN/TAP 为用户空间程序提供数据包的接收和传输。可以将其视为简单的点对点或以太网设备。它们不是从物理介质接收数据包，而是从用户空间程序接收数据包。它们也不是通过物理介质发送数据包，而是将其写入用户空间程序。

也就是说，TUN/TAP 接口是没有关联物理设备的虚拟接口。用户空间程序可以附加到 TUN/TAP 接口并处理发送到该接口的流量，如图 5-5 所示。

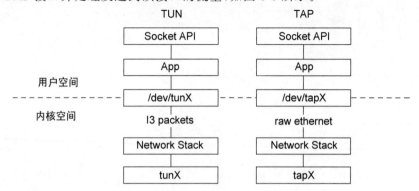

图 5-5　Linux 中 TUN/TAP 虚拟设备工作原理

使用 TUN/TAP 技术可以在主机上构建虚拟网络接口，它们的功能类似于物理网卡，可以为其分配 IP、将数据包路由到该接口、分析流量等。TUN/TAP 有两种常用的应用场景：VPN 和云计算。

TUN 与 TAP 有什么区别呢？

TUN(tunnel)设备在 OSI 模型的第 3 层运行，它实现的是虚拟的 IP 点对点接口，应用程序只能从此接口发送或接收 IP 数据包。它没有 MAC 地址，而且不能成为网桥的 slave 设备。

TAP(Network Tap)的运行方式与 TUN 极为相似，但是它运行在 OSI 模型的第 2 层上，应用程序从此接口发送和接收原始以太网数据包，所以它与以太网卡特别相似。它有 MAC 地址，可以成为网桥的 slave 设备。

下面通过实验讲解 TAP 设备的管理。

首先是 ip 命令。ip 是 iproute 软件包里面的一个强大的网络配置工具，其中 tuntap 子命令可用于管理 TUN/TAP 接口，命令如下：

```
1 # ip link
  1: lo: < LOOPBACK,UP,LOWER_UP > mtu 65536 qdisc noqueue state UNKNOWN mode DEFAULT group default qlen 1000
     link/loopback 00:00:00:00:00:00 brd 00:00:00:00:00:00
  2: ens32: < BROADCAST,MULTICAST,UP,LOWER_UP > mtu 1500 qdisc fq_codel state UP mode DEFAULT group default qlen 1000
     link/ether 00:0c:29:f7:6b:c8 brd ff:ff:ff:ff:ff:ff
  3: virbr0: < BROADCAST,MULTICAST,UP,LOWER_UP > mtu 1500 qdisc noqueue state UP mode DEFAULT group default qlen 1000
     link/ether 52:54:00:e0:41:ac brd ff:ff:ff:ff:ff:ff
  4: virbr0-nic: < BROADCAST,MULTICAST > mtu 1500 qdisc fq_codel master virbr0 state DOWN mode DEFAULT group default qlen 1000
     link/ether 52:54:00:e0:41:ac brd ff:ff:ff:ff:ff:ff
  6: vnet0: < BROADCAST,MULTICAST,UP,LOWER_UP > mtu 1500 qdisc fq_codel master virbr0 state UNKNOWN mode DEFAULT group default qlen 1000
     link/ether fe:54:00:27:5f:c9 brd ff:ff:ff:ff:ff:ff
  7: vnet1: < BROADCAST,MULTICAST,UP,LOWER_UP > mtu 1500 qdisc fq_codel master virbr0 state UNKNOWN mode DEFAULT group default qlen 1000
     link/ether fe:54:00:9d:57:9c brd ff:ff:ff:ff:ff:ff

2 # ip tuntap help
  Usage: ip tuntap { add | del | show | list | lst | help } [ dev PHYS_DEV ]
          [ mode { tun | tap } ] [ user USER ] [ group GROUP ]
          [ one_queue ] [ pi ] [ vnet_hdr ] [ multi_queue ] [ name NAME ]

  Where: USER := { STRING | NUMBER }
         GROUP := { STRING | NUMBER }
```

```
3 # ip tuntap list
  virbr0-nic: tap persist
  vnet0: tap vnet_hdr
  vnet1: tap
```

第 1 行 ip link 命令用于显示当前主机上的所有物理及虚拟的网络设备。

第 2 行命令用于显示 ip 命令的 tuntap 子命令的帮助。

第 3 行命令用于显示当前的 TUN/TAP 设备信息,此处共有 3 个 TAP 设备。

提示：libvirt 在创建 TAP 设备时,会根据虚拟机网卡类型及用途为 TAP 设备增加额外的标记(Flag),例如为 virtio 类型网卡增加的标记是 vnet_hdr,而对于传统的 e1000 的网卡则不设置标记。

接下来添加 1 个 TAP 设备,然后查看这个新设备,命令如下：

```
4 # ip tuntap add dev tap-nic1 mode tap

5 # ip tuntap list
  virbr0-nic: tap persist
  vnet0: tap vnet_hdr
  vnet1: tap
  tap-nic1: tap persist

6 # ip link
  1: lo: <LOOPBACK,UP,LOWER_UP> mtu 65536 qdisc noqueue state UNKNOWN mode DEFAULT group default qlen 1000
      link/loopback 00:00:00:00:00:00 brd 00:00:00:00:00:00
  2: ens32: <BROADCAST,MULTICAST,UP,LOWER_UP> mtu 1500 qdisc fq_codel state UP mode DEFAULT group default qlen 1000
      link/ether 00:0c:29:f7:6b:c8 brd ff:ff:ff:ff:ff:ff
  3: virbr0: <BROADCAST,MULTICAST,UP,LOWER_UP> mtu 1500 qdisc noqueue state UP mode DEFAULT group default qlen 1000
      link/ether 52:54:00:e0:41:ac brd ff:ff:ff:ff:ff:ff
  4: virbr0-nic: <BROADCAST,MULTICAST> mtu 1500 qdisc fq_codel master virbr0 state DOWN mode DEFAULT group default qlen 1000
      link/ether 52:54:00:e0:41:ac brd ff:ff:ff:ff:ff:ff
  6: vnet0: <BROADCAST,MULTICAST,UP,LOWER_UP> mtu 1500 qdisc fq_codel master virbr0 state UNKNOWN mode DEFAULT group default qlen 1000
      link/ether fe:54:00:27:5f:c9 brd ff:ff:ff:ff:ff:ff
  7: vnet1: <BROADCAST,MULTICAST,UP,LOWER_UP> mtu 1500 qdisc fq_codel master virbr0 state UNKNOWN mode DEFAULT group default qlen 1000
      link/ether fe:54:00:9d:57:9c brd ff:ff:ff:ff:ff:ff
  8: tap-nic1: <BROADCAST,MULTICAST> mtu 1500 qdisc noop state DOWN mode DEFAULT group default qlen 1000
```

```
    link/ether 76:0a:b9:fa:83:e9 brd ff:ff:ff:ff:ff:ff

7 # ip addr add 172.16.123.1/24 dev tap-nic1
```

第 4 行命令添加了 1 个名为 tap-nic1 的 TAP 设备。从第 5、6 行命令的输出可以看到这个新的设备。

我们可以像管理物理网卡一样管理虚拟网络接口,例如第 7 行命令给它设置了 1 个静态的 IP 地址。接下来删除这个 TAP 设备,然后查看是否成功删除,命令如下:

```
8 # ip tuntap del dev tap-nic1 mode tap

9 # ip tuntap list
  virbr0-nic: tap persist
  vnet0: tap vnet_hdr
  vnet1: tap
```

实验做完了,第 8 行命令用于删除这个 TAP 设备,第 9 行命令用于查看 TAP 设备列表。

RHEL/CentOS 8 中使用 ip 命令配置设备信息,大部分会在设备重启后还原。如果需要持久的 TUN/TAP 设置,则推荐使用 NetworkManager 来创建。下面我们通过 NetworkManager 中的命令行工具 nmcli 来做类似的实验,命令如下:

```
10 # nmcli connection add type tun con-name newcon1 ifname tap-nic2 mode tap ip4 172.16.123.2/24
   Connection 'newcon1' (09df2d02-dfc1-4a92-9d24-72dd97fe0756) successfully added.

11 # nmcli connection
   NAME      UUID                                    TYPE      DEVICE
   ens32     152beb06-47c5-c5e8-95a9-385590654382    ethernet  ens32
   newcon1   09df2d02-dfc1-4a92-9d24-72dd97fe0756    tun       tap-nic2
   virbr0    46edfc4d-8edb-4560-8a86-c84cad9b61c9    bridge    virbr0
   vnet0     540308fa-9ffa-40ec-8860-dd5ce13dde73    tun       vnet0
   vnet1     7a8486a7-88d0-4423-950a-0705d7aa3025    tun       vnet1

12 # nmcli device
   DEVICE      TYPE      STATE      CONNECTION
   ens32       ethernet  connected  ens32
   tap-nic2    tun       connected  newcon1
   virbr0      bridge    connected  virbr0
   vnet0       tun       connected  vnet0
   vnet1       tun       connected  vnet1
   lo          loopback  unmanaged  --
   virbr0-nic  tun       unmanaged  --
```

NetworkManager 将所有网络配置存储为连接(connection),它们是描述如何创建或描述如何连接到网络的配置(2 层详细信息、IP 地址等)的集合。NetworkManager 支持的连接类型有多种,包括 ethernet、wifi、pppoe、infiniband、bluetooth、bond、bridge、tun、vxlan 等。需要注意的是,NetworkManager 不区分连接类型究竟是 TUN 还是 TAP,统一称为 TUN。

第 10 行命令创建了一个新的连接,将 type 设置为 tun,通过 con-name 指定了连接的名称为 newcon1,通过 ifname 指定了接口名称为 tap-nic2,通过 mode 指定了模式为 tap,通过 ip4 指定了 IP 地址为 172.16.123.2/24。

从第 11 行命令的输出中可以看到这个名为 newcon1 的新连接及其属性。

从第 12 行命令的输出中可以看到新设备 tap-nic2,type 显示为 tun。我们可以认为这是指连接的类型,而且不是设备的类型。tap-nic2 的类型还是 tap,这个可以从第 14 行命令的输出中看出,命令如下:

```
13 # ip address show tap-nic2
    7: tap-nic2: <NO-CARRIER,BROADCAST,MULTICAST,UP> mtu 1500 qdisc fq_codel state DOWN group default qlen 1000
        link/ether ce:ba:06:00:66:ef brd ff:ff:ff:ff:ff:ff
        inet 172.16.123.2/24 brd 172.16.123.255 scope global noprefixroute tap-nic2
           valid_lft forever preferred_lft forever
        inet6 fe80::c5d3:c6b8:d79:81a7/64 scope link tentative noprefixroute
           valid_lft forever preferred_lft forever

14 # ip tuntap
    virbr0-nic: tap persist
    vnet0: tap vnet_hdr
    vnet1: tap
    tap-nic2: tap persist

15 # nmcli connection delete newcon1
    Connection 'newcon1' (09df2d02-dfc1-4a92-9d24-72dd97fe0756) successfully deleted.

16 # ip tuntap list
    virbr0-nic: tap persist
    vnet0: tap vnet_hdr
    vnet1: tap
```

从第 13 行命令的输出中可看到虚拟网络接口 tap-nic2 的信息,包括随机生成的 MAC 地址、IPv4 地址、IPv6 地址等信息。

第 15 行命令会清除连接 newcon1,这个操作也会同时删除虚拟网络接口设备 tap-nic2。

5.3 网桥工作原理与管理

网桥是一种数据链路层的设备。2.4 版以后的 Linux 内核之中已经集成了网桥功能。在虚拟化解决方案中，网桥是一个很重要的组件。

5.3.1 考察现有网桥

在 RHEL/CentOS 8 中，可以通过 iproute、NetworkManager 和 Cockpit 来查看及了解当前主机上的网桥情况。

在"5.1.1 查看宿主机的网络环境"一节中讲解了如何通过 iproute 和 NetworkManager 来查看宿主机当前的网桥，下面我们通过 Cockpit 查看这部分配置。

在 Cockpit 中单击 Networking 就可以看到宿主机的所有网络接口信息，在其中可以看到网桥 virbr0，如图 5-6 所示。

图 5-6　Cockpit 中的网络接口

单击 virbr0 就可以看到这个网桥的详细信息，包括 IP 地址、支持 STP、转发延迟及两个端口 vnet0 和 vnet1，如图 5-7 所示。

提示：RHEL 7.7 已经不推荐使用 bridge-utils 了，所以在 RHEL 8/CentOS 8 中已经无法使用 brctl 命令了，我们完全可以使用 iproute、NetworkManager 和 Cockpit 来替代它。

5.3.2 通过 iproute 管理网桥

在 iproute 软件包中，ip 命令是最常用的命令，下面通过它创建一个网桥接口，网桥接

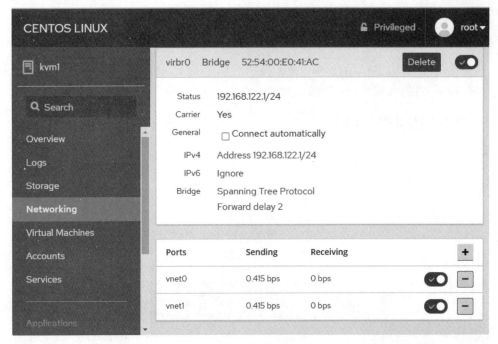

图 5-7　Cockpit 中的网桥详细信息

口的名称可以用于表示网桥,命令如下:

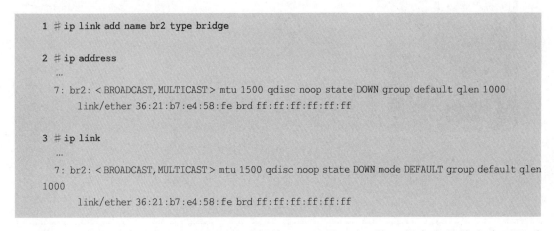

第 1 行命令创建了 1 个名为 br2 的网桥接口。从第 2 行、第 3 行命令的输出中可以看到这个新的网桥。网桥都有 1 个随机生成的 MAC 地址,注意此时这个新网桥的 MAC 地址是 36:21:b7:e4:58:fe。

有了网桥,下面就可以为其添加子接口了。网桥上的子接口既可以是物理接口也可以是虚拟接口,本次实验将使用虚拟接口,命令如下:

```
4 # ip tuntap add dev tap-nic1 mode tap

5 # ip tuntap add dev tun-nic1 mode tun

6 # ip tuntap
  virbr0-nic: tap persist
  vnet0: tap vnet_hdr
  vnet1: tap
  tap-nic1: tap persist
  tun-nic1: tun persist

7 # ip link
  ...
  7: br2: <BROADCAST,MULTICAST> mtu 1500 qdisc noop state DOWN mode DEFAULT group default qlen 1000
      link/ether 36:21:b7:e4:58:fe brd ff:ff:ff:ff:ff:ff
  8: tap-nic1: <BROADCAST,MULTICAST> mtu 1500 qdisc noop state DOWN mode DEFAULT group default qlen 1000
      link/ether 26:3a:41:e8:7f:78 brd ff:ff:ff:ff:ff:ff
  9: tun-nic1: <POINTOPOINT,MULTICAST,NOARP> mtu 1500 qdisc noop state DOWN mode DEFAULT group default qlen 500
      link/none
```

第 4 行命令添加了 1 个名为 tap-nic1 的 TAP 类型的虚拟接口。出于对比的目的，使用第 5 行命令再创建了 1 个名为 tun-nic1 的 TUN 类型的虚拟接口。可以在第 6 行命令的输出中看到这 2 个虚拟接口。

从第 7 行命令的输出中可以看出，tap-nic1 与 br2 的 MAC 地址是不同的。由于 tun-nic1 是一个 3 层的设备，所以它是没有 MAC 地址的。

接下来激活连接及设置子接口，命令如下：

```
8 # ip link set br2 up

9 # ip link set tap-nic1 up

10 # ip link set tun-nic1 up

11 # ip link set tap-nic1 master br2

12 # ip link set tun-nic1 master br2
   RTNETLINK answers: Invalid argument
```

ip 命令创建的连接默认为没有被激活，所以第 8、9、10 行命令激活了这 3 个连接。第 11 行命令将这个名为 tap-nic1 的 TAP 设备设置为网桥 br2 的子接口，命令没有输出表示成功。由于 tun-nic1 是一个 3 层的设备，它无法充当网桥的子接口，所以第 12 行命令会返回一个错误。

查看网络配置,命令如下:

```
13 # ip addr
    …
    3: virbr0: <BROADCAST,MULTICAST,UP,LOWER_UP> mtu 1500 qdisc noqueue state UP group default qlen 1000
        link/ether 52:54:00:e0:41:ac brd ff:ff:ff:ff:ff:ff
        inet 192.168.122.1/24 brd 192.168.122.255 scope global virbr0
           valid_lft forever preferred_lft forever
    4: virbr0-nic: <BROADCAST,MULTICAST> mtu 1500 qdisc fq_codel master virbr0 state DOWN group default qlen 1000
        link/ether 52:54:00:e0:41:ac brd ff:ff:ff:ff:ff:ff
    …
    7: br2: <NO-CARRIER,BROADCAST,MULTICAST,UP> mtu 1500 qdisc noqueue state DOWN group default qlen 1000
        link/ether 26:3a:41:e8:7f:78 brd ff:ff:ff:ff:ff:ff
        inet6 fe80::3421:b7ff:fee4:58fe/64 scope link
           valid_lft forever preferred_lft forever
    8: tap-nic1: <NO-CARRIER,BROADCAST,MULTICAST,UP> mtu 1500 qdisc fq_codel master br2 state DOWN group default qlen 1000
        link/ether 26:3a:41:e8:7f:78 brd ff:ff:ff:ff:ff:ff
    9: tun-nic1: <NO-CARRIER,POINTOPOINT,MULTICAST,NOARP,UP> mtu 1500 qdisc fq_codel state DOWN group default qlen 500
        link/none
```

我们注意一下第 13 命令的输出:tap-nic1 的 master 是网桥 br2,而且网桥 br2 会将它的第 1 个接口的 MAC 地址当作自己的 MAC,也就是 tap-nic1 的 MAC 地址 26:3a:41:e8:7f:78。这是 Linux 网桥一个特性:它会自动将第 1 个网络接口的 MAC 当作自己的 MAC 地址,所以 libvirt 在创建网桥 virbr0 时,会先创建一个 TAP 类型的接口 virbr0-nic,这样 virbr0 和 virbr0-nic 的 MAC 地址都是 52:54:00:e0:41:ac。

提示:默认情况下 libvirt 生成的 MAC 地址是 52:54:00 开头的。

在 iproute 软件包中,除了 ip 命令外,还有 1 个名为 bridge 的命令也与网桥有关,命令如下:

```
14 # bridge link
    4: virbr0-nic: <BROADCAST,MULTICAST> mtu 1500 master virbr0 state disabled priority 32 cost 100
    5: vnet0: <BROADCAST,MULTICAST,UP,LOWER_UP> mtu 1500 master virbr0 state forwarding priority 32 cost 100
    6: vnet1: <BROADCAST,MULTICAST,UP,LOWER_UP> mtu 1500 master virbr0 state forwarding priority 32 cost 100
```

```
    8: tap-nic1: <NO-CARRIER,BROADCAST,MULTICAST,UP> mtu 1500 master br2 state disabled
priority 32 cost 100
```

做完实验之后,我们可以使用下述命令清理实验环境。

```
15 # ip link set tap-nic1 nomaster

16 # ip link delete br2 type bridge

17 # ip tuntap delete tap-nic1 mode tap

18 # ip tuntap delete tun-nic1 mode tun
```

5.3.3　通过 NetworkManager 管理网桥

可以通过 nmcli 的 connection 子命令来管理网桥。与 ip 命令类似,它也是先创建网桥的连接再创建子接口。不同之处是它会自动创建网桥接口文件。

在下面的实验中,我们将使用一块新网卡 ens34 作为新网桥的子接口,代码如下:

```
1 # nmcli connection
  NAME      UUID                                    TYPE      DEVICE
  ens32     152beb06-47c5-c5e8-95a9-385590654382    ethernet  ens32
  ens34     293e8ad9-d491-46ee-a0ac-789fc12c5407    ethernet  ens34
  virbr0    b4cc48df-f27f-4c97-9709-bcca97a890e5    bridge    virbr0
  vnet0     39e2fcf8-26f4-4e4f-be22-1eb79f43c01b    tun       vnet0
  vnet1     84718c08-1d46-4d0b-b80c-79fb448eccce    tun       vnet1

2 # nmcli device
  DEVICE       TYPE       STATE       CONNECTION
  ens32        ethernet   connected   ens32
  ens34        ethernet   connected   ens34
  virbr0       bridge     connected   virbr0
  vnet0        tun        connected   vnet0
  vnet1        tun        connected   vnet1
  lo           loopback   unmanaged   --
  virbr0-nic   tun        unmanaged   --

4 # ls /etc/sysconfig/network-scripts/
  ifcfg-ens32 ifcfg-ens34

5 # cat /etc/sysconfig/network-scripts/ifcfg-ens34
  TYPE=Ethernet
  PROXY_METHOD=none
  BROWSER_ONLY=no
```

```
    BOOTPROTO = dhcp
    DEFROUTE = yes
    IPV4_FAILURE_FATAL = no
    IPV6INIT = yes
    IPV6_AUTOCONF = yes
    IPV6_DEFROUTE = yes
    IPV6_FAILURE_FATAL = no
    IPV6_ADDR_GEN_MODE = stable-privacy
    NAME = ens34
    UUID = 293e8ad9-d491-46ee-a0ac-789fc12c5407
    DEVICE = ens34
    ONBOOT = yes

6 # ip address
    …
    3: ens34: <BROADCAST,MULTICAST,UP,LOWER_UP> mtu 1500 qdisc fq_codel state UP group default qlen 1000
        link/ether 00:0c:29:f7:6b:d2 brd ff:ff:ff:ff:ff:ff
        inet 192.168.114.130/24 brd 192.168.114.255 scope global dynamic noprefixroute ens34
            valid_lft 1750sec preferred_lft 1750sec
        inet6 fe80::697b:329c:b7b3:238d/64 scope link noprefixroute
            valid_lft forever preferred_lft forever
    …
```

从第 1 行及第 2 行命令的输出中可以看出：有一个名为 ens34 的连接，它有一个同名的以太网接口设备。

在 /etc/sysconfig/network-scripts/ 目录中有一个名为 ifcfg-ens34 的网桥接口文件，从第 5 行的命令输出就可以看到这个配置文件的内容。系统初始化时会自动根据这个配置文件进行相应的设置，例如通过 DHCP 给 ens34 分配的 IP 地址，这可以从第 6 行命令的输出中看到 IP 地址是 192.168.114.130/24。

提示：如果 /etc/sysconfig/network-scripts/ 目录中没有新网卡的接口文件，则除了手工创建之外，还可以通过 nmcli device connect ens34 命令来生成一个新的接口文件。

接下来执行以下命令：

```
7 # nmcli connection add type bridge autoconnect yes con-name virbr1 ifname virbr1
    Connection 'virbr1' (24fc8f00-ca86-47da-817c-cb392c4d15b4) successfully added.

8 # nmcli connection
    NAME    UUID                                    TYPE      DEVICE
    virbr1  24fc8f00-ca86-47da-817c-cb392c4d15b4    bridge    virbr1
    ens32   152beb06-47c5-c5e8-95a9-385590654382    ethernet  ens32
```

```
ens34       861f3757-4b5b-464a-9d56-fd7c0b4e8cee    ethernet   ens34
virbr0      d767930b-ee30-49ee-8d59-e51708f1cc1d    bridge     virbr0

9 # nmcli device
DEVICE        TYPE        STATE                                    CONNECTION
ens32         ethernet    connected                                ens32
ens34         ethernet    connected                                ens34
virbr0        bridge      connected                                virbr0
virbr1        bridge      connecting (getting IP configuration)    virbr1
lo            loopback    unmanaged                                --
virbr0-nic    tun         unmanaged                                --

10 # ls /etc/sysconfig/network-scripts/
ifcfg-ens32  ifcfg-ens34  ifcfg-virbr1

11 # cat /etc/sysconfig/network-scripts/ifcfg-virbr1
STP=yes
BRIDGING_OPTS=priority=32768
TYPE=Bridge
PROXY_METHOD=none
BROWSER_ONLY=no
BOOTPROTO=dhcp
DEFROUTE=yes
IPV4_FAILURE_FATAL=no
IPV6INIT=yes
IPV6_AUTOCONF=yes
IPV6_DEFROUTE=yes
IPV6_FAILURE_FATAL=no
IPV6_ADDR_GEN_MODE=stable-privacy
NAME=virbr1
UUID=24fc8f00-ca86-47da-817c-cb392c4d15b4
DEVICE=virbr1
ONBOOT=yes
```

第 7 行命令成功添加了一个新的连接并设置了属性，type 指定的类型是 bridge，autoconnect 指定为自动启动，con-name 指定的连接名称为 virbr1，ifname 指定的接口名称也是 virbr1。

从第 8 行、第 9 行的输出中可以看到这个网桥的连接及设备信息。

第 7 行命令会在 /etc/sysconfig/network-scripts/ 目录中生成一个名为 ifcfg-virbr1 的网络接口文件。

注意第 11 行命令的输出中的两个属性：TYPE=Bridge 和 BOOTPROTO=dhcp。

接下来执行的命令如下：

```
12 # ip address
    ...
    3: ens34: <BROADCAST,MULTICAST,UP,LOWER_UP> mtu 1500 qdisc fq_codel state UP group default qlen 1000
        link/ether 00:0c:29:f7:6b:d2 brd ff:ff:ff:ff:ff:ff
        inet 192.168.114.130/24 brd 192.168.114.255 scope global dynamic noprefixroute ens34
           valid_lft 1582sec preferred_lft 1582sec
        inet6 fe80::697b:329c:b7b3:238d/64 scope link noprefixroute
           valid_lft forever preferred_lft forever
    ...
    6: virbr1: <NO-CARRIER,BROADCAST,MULTICAST,UP> mtu 1500 qdisc noqueue state DOWN group default qlen 1000
        link/ether 92:23:18:cb:00:e3 brd ff:ff:ff:ff:ff:ff

13 # nmcli connection modify virbr1 ipv4.addresses 172.16.123.11/24 ipv4.method manual

14 # cat /etc/sysconfig/network-scripts/ifcfg-virbr1
    STP=yes
    BRIDGING_OPTS=priority=32768
    TYPE=Bridge
    PROXY_METHOD=none
    BROWSER_ONLY=no
    BOOTPROTO=none
    DEFROUTE=yes
    IPV4_FAILURE_FATAL=no
    IPV6INIT=yes
    IPV6_AUTOCONF=yes
    IPV6_DEFROUTE=yes
    IPV6_FAILURE_FATAL=no
    IPV6_ADDR_GEN_MODE=stable-privacy
    NAME=virbr1
    UUID=24fc8f00-ca86-47da-817c-cb392c4d15b4
    DEVICE=virbr1
    ONBOOT=yes
    IPADDR=172.16.123.11
    PREFIX=24
```

从第 12 行命令的输出可以看出：由于还没有激活新网桥 virbr1，所以它没有从 DHCP 服务器获得 IP 地址。需要注意的是 virbr1 的 MAC 地址是随机生成的 92:23:18:cb:00:e3，与 ens34 的 MAC 地址 00:0c:29:f7:6b:d2 并不相同。

第 13 行命令将 virbr1 的 IP 地址修改为静态 IP 地址 172.16.123.11/24。这个修改会反映在 /etc/sysconfig/network-scripts/ 目录中的网格接口文件 ifcfg-virbr1 的变化上。

接下来执行的命令如下：

```
15 #nmcli connection delete ens34
   Connection 'ens34' (861f3757-4b5b-464a-9d56-fd7c0b4e8cee) successfully deleted.

16 #nmcli connection add type bridge-slave autoconnect yes con-name ens34 master virbr1
   Connection 'ens34' (ad4ae2e5-f95f-415a-a9e1-23e2ff650956) successfully added.

17 #nmcli connection
   NAME    UUID                                   TYPE      DEVICE
   virbr1  24fc8f00-ca86-47da-817c-cb392c4d15b4   bridge    virbr1
   ens32   152beb06-47c5-c5e8-95a9-385590654382   ethernet  ens32
   virbr0  d767930b-ee30-49ee-8d59-e51708f1cc1d   bridge    virbr0
   ens34   ad4ae2e5-f95f-415a-a9e1-23e2ff650956   ethernet  ens34

18 #nmcli device
   DEVICE       TYPE       STATE      CONNECTION
   ens32        ethernet   connected  ens32
   virbr1       bridge     connected  virbr1
   virbr0       bridge     connected  virbr0
   ens34        ethernet   connected  ens34
   lo           loopback   unmanaged  --
   virbr0-nic   tun        unmanaged  --

19 #ls /etc/sysconfig/network-scripts/
   ifcfg-ens32  ifcfg-ens34  ifcfg-virbr1

20 #cat /etc/sysconfig/network-scripts/ifcfg-ens34
   TYPE=Ethernet
   NAME=ens34
   UUID=ad4ae2e5-f95f-415a-a9e1-23e2ff650956
   ONBOOT=yes
   BRIDGE=virbr1

21 #nmcli connection up virbr1
   Connection successfully activated (master waiting for slaves) (D-Bus active path: /org/freedesktop/NetworkManager/ActiveConnection/7)
```

网络设备 ens34 现在还属于一个同名的连接，所以需要先将这个连接删除。第 15 行命令将连接 ens34 删除。

第 16 行命令将网络设备 ens34 添加到网桥 virbr1，从而成为网桥的子接口。type 指定的类型为 bridge-slave，autoconnect 指定为自动启动，con-name 指定的连接名称为 ens34，master 指定的网桥名称是 virbr1。

从第 17 行、第 18 行的输出中可以看到 virbr1、ens34 的连接及设备信息。

第16行命令会为ens34在/etc/sysconfig/network-scripts/目录中生成一个新的网络接口文件ifcfg-ens34。通过第20行命令查看这个新文件,新文件很简单,核心属性是BRIDGE=virbr1,它为ens34指定了master。

第21行命令激活了网桥。在RHEL/CentOS 8中,这些操作会在系统的/var/log/messages生成日志条目。当操作失败时,阅读这些日志条件将有助于我们进行排错操作。

接下来执行的命令如下:

```
22 # ip address
   ...
   3: ens34: <BROADCAST,MULTICAST,UP,LOWER_UP> mtu 1500 qdisc fq_codel master virbr1 state UP group default qlen 1000
      link/ether 00:0c:29:f7:6b:d2 brd ff:ff:ff:ff:ff:ff
   ...
   6: virbr1: <BROADCAST,MULTICAST,UP,LOWER_UP> mtu 1500 qdisc noqueue state UP group default qlen 1000
      link/ether 00:0c:29:f7:6b:d2 brd ff:ff:ff:ff:ff:ff
      inet 172.16.123.11/24 brd 172.16.123.255 scope global noprefixroute virbr1
         valid_lft forever preferred_lft forever
      inet6 fe80::e708:9d2e:c0ca:f913/64 scope link noprefixroute
         valid_lft forever preferred_lft forever

23 # bridge link
   3: ens34: <BROADCAST,MULTICAST,UP,LOWER_UP> mtu 1500 master virbr1 state forwarding priority 32 cost 100
   5: virbr0-nic: <BROADCAST,MULTICAST> mtu 1500 master virbr0 state disabled priority 32 cost 100
```

从第22行命令的输出可以看出:ens34的master是virbr1,这说明它是网桥virbr1的子接口。同时要注意virbr1的MAC地址已经不再是原有那个随机成的MAC地址了,而是借用了第1个子接口,即ens34的MAC地址00:0c:29:f7:6b:d2。

第23行命令的输出也显示出ens34的master是virbr1。

创建网桥的实验做完了,下面我们进行删除及恢复操作,命令如下:

```
24 # mkdir ~/bak

25 # cp /etc/sysconfig/network-scripts/ifcfg-* ~/bak

26 # nmcli connection down virbr1
   Connection 'virbr1' successfully deactivated (D-Bus active path: /org/freedesktop/NetworkManager/ActiveConnection/7)

27 # nmcli connection delete virbr1
   Connection 'virbr1' (24fc8f00-ca86-47da-817c-cb392c4d15b4) successfully deleted.
```

```
28 #nmcli connection delete ens34
   Connection 'ens34' (ad4ae2e5-f95f-415a-a9e1-23e2ff650956) successfully deleted.

29 #nmcli device connect ens34
   Device 'ens34' successfully activated with '6d322e2c-af9c-4c87-9740-d4cf6e076bfc'.
```

为了后续的实验,使用第 24 行、第 25 行命令将当前的网桥接口文件备份到~/bak 目录中。

使用第 26 行命令关闭 virbr1,然后删除连接 virbr1、ens34。

通过第 29 行命令会重新连接设备 ens34,并在/etc/sysconfig/network-scripts/目录生成包含默认属性值的 ifcfg-ens34 文件。

提示:在 NetworkManager 中还有图形化网络连接编辑器 nm-connection-editor。与 nmclic 类似,它也可以很方便地添加、删除和修改 NetworkManager 存储的网络连接。

5.3.4 通过网络接口文件管理网桥

在目录/etc/sysconfig/network-scripts/下创建网桥的配置文件 ifcfg-virbr1 和子接口配置文件 ifcfg-ens34,也可以很方便地创建网桥,命令如下:

```
1 #vi /etc/sysconfig/network-scripts/ifcfg-virbr1
  #添加以下内容:
  TYPE = Bridge
  STP = yes
  BOOTPROTO = none
  NAME = virbr1
  DEVICE = virbr1
  ONBOOT = yes
  IPADDR = 172.16.123.11
  PREFIX = 24

2 #vi /etc/sysconfig/network-scripts/ifcfg-ens34
  #添加以下内容:
  TYPE = Ethernet
  NAME = ens34
  ONBOOT = yes
  BRIDGE = virbr1

3 #nmcli connection reload

4 #ip address
  ...
```

```
    3: ens34: < BROADCAST,MULTICAST,UP,LOWER_UP > mtu 1500 qdisc fq_codel master virbr1 state UP
group default qlen 1000
        link/ether 00:0c:29:f7:6b:d2 brd ff:ff:ff:ff:ff:ff
    …
    6: virbr1: < BROADCAST,MULTICAST,UP,LOWER_UP > mtu 1500 qdisc noqueue state UP group default
qlen 1000
        link/ether 00:0c:29:f7:6b:d2 brd ff:ff:ff:ff:ff:ff
        inet 172.16.123.11/24 brd 172.16.123.255 scope global noprefixroute virbr1
            valid_lft forever preferred_lft forever
        inet6 fe80::20c:29ff:feb3:7340/64 scope link
            valid_lft forever preferred_lft forever
```

如第 1 行、第 2 行命令所示，不管是网桥的配置文件还是子接口的配置文件，仅需要必需的属性即可。当然，还可以将上一实验备份在目录 ~/bak/ 下的 ifcfg-virbr1 和 ifcfg-ens34 直接复制到目录 /etc/sysconfig/network-scripts/ 下。

当执行第 3 行命令重新加载网络连接配置之后，网桥就配置成功了。这可以从第 4 行命令的输出信息中得到验证。如果失败，则可以查看系统日志 /var/log/messages 中的条目来查找错误原因。

```
5 # rm /etc/sysconfig/network-scripts/ifcfg-virbr1

6 # rm /etc/sysconfig/network-scripts/ifcfg-ens34

7 # nmcli connection down virbr1
  Connection 'virbr1' successfully deactivated (D-Bus active path: /org/freedesktop/
NetworkManager/ActiveConnection/6)

8 # nmcli connection reload
```

清除网桥 virbr1 也很简单，首先执行第 5 行、第 6 行命令删除网桥接口配置文件，然后在关闭网桥 virbr1 之后重新加载网络连接即可。

5.3.5 通过 Cockpit 管理网桥

相对于前面几种方法，使用 Cockpit 来管理网桥是最简单的。在网络信息中单击 Add Bridge 就可以添加新网桥了，如图 5-8 所示。

输入新网桥名称 virbr1，选中子接口（Cockpit 将其称为 port）ens34，其他选项保持默认，然后单击 Apply 按钮创建新网桥，如图 5-9 所示。

通过 Cockpit 来创建新网桥 virbr1，会自动在 /etc/sysconfig/network-scripts/ 目录中创建配置文件 ifcfg-ens34 和 ifcfg-virbr1。

Cockpit 默认使用 DHCP 服务向新网桥分配 IP 地址，当然也可以单击新网桥 virbr1 的链接进行修改，如图 5-10 所示。

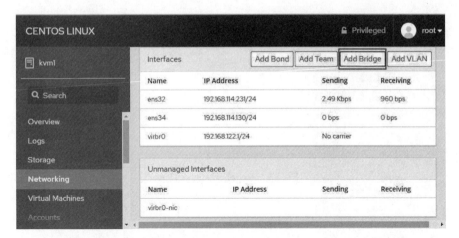

图 5-8　Cockpit 中的网络接口

图 5-9　在 Cockpit 中创建新的网桥

可以在新页面中设置 IP 地址、STP 及子接口等信息，也可以单击 Delete 按钮删除网桥，如图 5-11 所示。

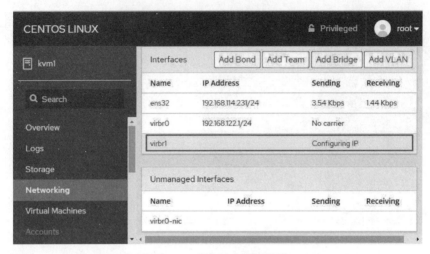

图 5-10 成功创建的新网桥

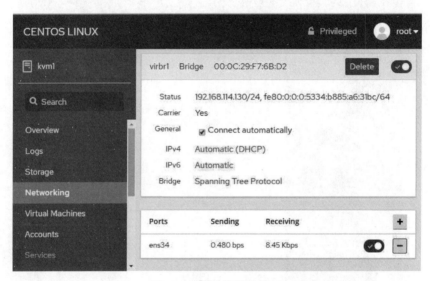

图 5-11 网桥的详细信息

5.4 KVM/libvirt 常用的网络类型

在掌握了 TUN/TAP 设备及网桥基本原理之后,我们就可以来学习 KVM/libvirt 常用的网络类型了。

本节先介绍虚拟机支持的网络、libvirt 管理的虚拟网络,然后讲解具体网络的工作原理。

5.4.1 虚拟机支持的网络

为虚拟机创建网络连接的时候，除了需要设置虚拟网卡的设备类型（如：virtio、e1000 或 rtl8139）之外，我们还需要为其指定网络连接的目标，可以将其分为两类。

（1）虚拟的网络：由 libvirt 来管理维护，例如 NAT、Routed 和 Isolated 等。虚拟机的配置是 interface type="network"。

（2）共享的物理设备：由宿主机操作系统管理维护，例如带物理网卡子接口的网桥、宿主机的物理网卡等。虚拟机配置是 interface type="bridge"或 interface type="direct"。

在 RHEL/CentOS 8 中，不同管理工具支持的虚拟机的网络类型、表示方法有些细微的差异。例如，virt-manager(virt-manager-2.2.1-3.el8.noarch)使用 Network source 的一个选项来设置虚拟机的网络连接，如图 5-12 所示。Cockpit(cockpit-machines-211.3-1.el8.noarch)则使用 Interface Type 和 Model 两个选项，如图 5-13 所示，而通过 virsh 的 edit 子命令编辑虚拟机的 XML 文件则可以支持所有的网络类型（https://libvirt.org/formatdomain.html）。

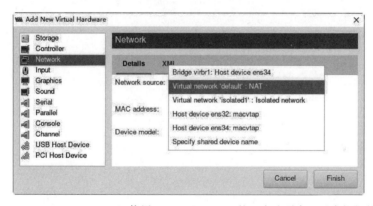

图 5-12　virt-manager-2.2.1-3.el8 使用 Network Source 的一个选项来配置虚拟机的网络接口

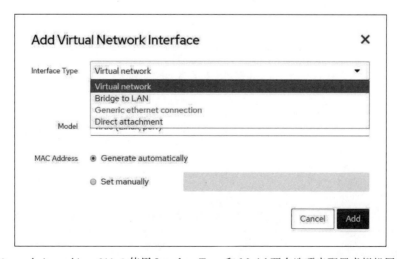

图 5-13　cockpit-machines-211.3 使用 Interface Type 和 Model 两个选项来配置虚拟机网络接口

表 5-1～表 5-8 是不同管理方式的对照。

表 5-1　连接到虚拟网络 default

工具	网卡配置
virt-manager	Virtual network 'default'：NAT
Cockpit	Interface Type：Virtual Network Source：default
虚拟机 XML	＜interface type＝"network"＞ 　＜source network＝"default"/＞ 　＜mac address＝"52:54:00:xx:xx:xx"/＞ 　＜model type＝"virtio"/＞ ＜/interface＞

表 5-2　连接到虚拟网络 isolated1

工具	网卡配置
virt-manager	Virtual Network 'isolated1'：Isolated Network
Cockpit	Interface Type：Virtual Network Source：isolated1
虚拟机 XML	＜interface type＝"network"＞ 　＜source network＝"isolated1"/＞ 　＜mac address＝"52:54:00:xx:xx:xx"/＞ 　＜model type＝"virtio"/＞ ＜/interface＞

表 5-3　连接到桥接网络 virbr1

工具	网卡配置
virt-manager	Bridge virbr1：Host device ens34
Cockpit	Interface Type：Virtual Network Source：virbr1
虚拟机 XML	＜interface type＝"bridge"＞ 　＜source bridge＝"virbr1"/＞ 　＜mac address＝"52:54:00:xx:xx:xx"/＞ 　＜model type＝"virtio"/＞ ＜/interface＞

表 5-4　连接宿主机设备 ens34（VEPA）

工具	网卡配置
virt-manager	Host device ens34：macvtap，Source mode：VEPA
Cockpit	Interface Type：Direct attachment Source：ens34 默认是 vepa
虚拟机 XML	＜interface type＝"direct"＞ 　＜source dev＝"ens34" mode＝"vepa"/＞ 　＜mac address＝"52:54:00:xx:xx:xx"/＞ 　＜model type＝"virtio"/＞ ＜/interface＞

表 5-5　连接宿主机设备 ens34（Bridge）

工具	网卡配置
virt-manager	Host device ens34：macvtap，Source mode：Bridge
Cockpit	Interface Type：Direct attachment Source：ens34 需要手工修改 mode

虚拟机 XML	`< interface type="direct">`
	　`< source dev="ens34" mode="bridge"/>`
	　`< mac address="52:54:00:xx:xx:xx"/>`
	　`< model type="virtio"/>`
	`</interface >`

表 5-6　连接宿主机设备 ens34（Private）

工具	网卡配置
virt-manager	Host device ens34：macvtap，Source mode：Private
Cockpit	Interface Type：Direct attachment Source：ens34 需要手工修改 mode

虚拟机 XML	`< interface type="direct">`
	　`< source dev="ens34" mode="private"/>`
	　`< mac address="52:54:00:xx:xx:xx"/>`
	　`< model type="virtio"/>`
	`</interface >`

表 5-7　连接宿主机设备 ens34（Passthrough）

工具	网卡配置
virt-manager	Host device ens34：macvtap，Source mode：Passthrough
Cockpit	Interface Type：Direct attachment Source：ens34 需要手工修改 mode

虚拟机 XML	`< interface type="direct">`
	　`< source dev="ens34" mode="passthrough"/>`
	　`< mac address="52:54:00:xx:xx:xx"/>`
	　`< model type="virtio"/>`
	`</interface >`

表 5-8　连接指定共享设备

工具	网卡配置
virt-manager	Specify shared device name Bridge name：brtest
Cockpit	不支持

虚拟机 XML	`< interface type="bridge">`
	　`< source bridge="brtest"/>`
	　`< mac address="52:54:00:xx:xx:xx"/>`
	　`< model type="virtio"/>`
	`</interface >`

5.4.2 libvirt 管理的虚拟网络

在 RHEL/CentOS 8 中启动 libvirtd 守护程序时，libvirtd 会读取/etc/libvirt/qemu/networks/autostart/中的符号链接所指向的虚拟网络配置文件，然后使用 NetworkManager 来创建相应的网桥、TAP 等设备，并根据需要配置 IP 转发、路由表和 iptables 中的 NAT 规则，从而为虚拟机提供虚拟网络环境。

对于虚拟网络，RHEL/CentOS 8 中不同管理工具也有一些细微的差异。例如：virt-manager（virt-manager-2.2.1-3.el8.noarch）使用的术语是 Mode 和 Forward to，共用 5 种 Mode，如图 5-14 所示。Cockpit（cockpit-machines-211.3-1.el8.noarch）使用的术语主要是 Forward Mode 和 Device，仅有 3 种 Forward Mode，如图 5-15 所示，而通过 virsh 的 net-create、net-define、net-edit 和 net-update 子命令编辑网络 XML 文件则支持所有的虚拟网络类型（https://libvirt.org/formatnetwork.html）。

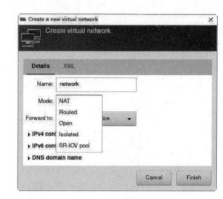

图 5-14　virt-manager-2.2.1-3.el8 创建虚拟网络

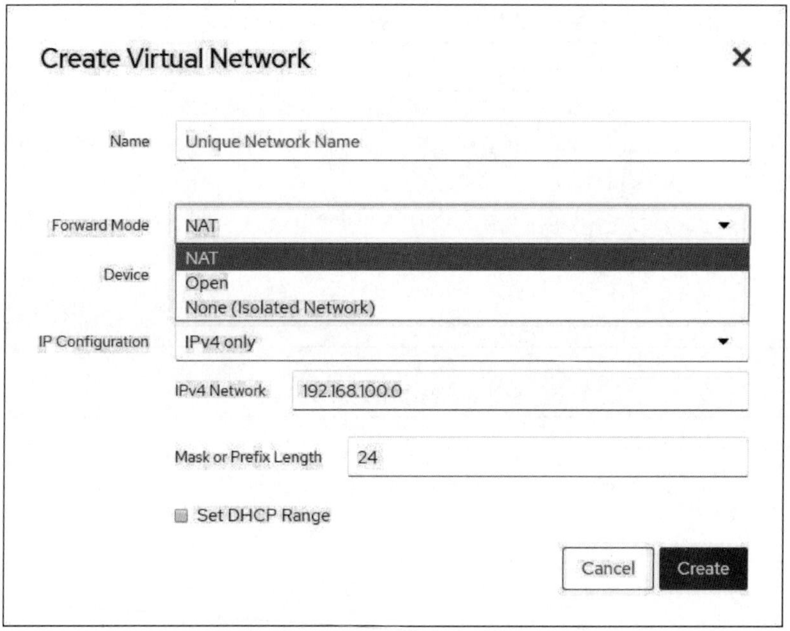

图 5-15　cockpit-machines-211.3 创建虚拟网络

表 5-9～表 5-13 是不同管理方式的对照。

表 5-9　NAT 模式虚拟网络

工具	虚拟网络配置
virt-manager	Mode：NAT，Forward to：Any physical device
Cockpit	Forward Mode：NAT，Device：Automatic
虚拟网络 XML	＜ network ＞
	＜ name ＞ nat1 ＜/name ＞
	＜ forward mode＝"nat"/＞
	＜ domain name＝"nat1"/＞
	＜ ip address＝"192.168.100.1" netmask＝"255.255.255.0"＞
	＜ dhcp ＞
	＜ range start＝"192.168.100.128" end＝"192.168.100.254"/＞
	＜/dhcp ＞
	＜/ip ＞
	＜/network ＞

表 5-10　开放模式虚拟网络

工具	虚拟网络配置
virt-manager	Mode：Open
Cockpit	Forward Mode：Open，IP Configuration：IPv4 only
虚拟网络 XML	＜ network ＞
	＜ name ＞ open1 ＜/name ＞
	＜ forward mode＝"open"/＞
	＜ domain name＝"open1"/＞
	＜ ip address＝"192.168.100.1" netmask＝"255.255.255.0"＞
	＜ dhcp ＞
	＜ range start＝"192.168.100.128" end＝"192.168.100.254"/＞
	＜/dhcp ＞
	＜/ip ＞
	＜/network ＞

表 5-11　隔离模式虚拟网络

工具	虚拟网络配置
virt-manager	Mode：Isolated
Cockpit	Forward Mode：None(Isolated Network)，IP Configuration：IPv4 only
虚拟网络 XML	＜ network ＞
	＜ name ＞ isolated1 ＜/name ＞
	＜ domain name＝"isolated1"/＞
	＜ ip address＝"192.168.100.1" netmask＝"255.255.255.0"＞

续表

工具	虚拟网络配置
虚拟网络 XML	< dhcp > < range start="192.168.100.128" end="192.168.100.254"/> </dhcp > </ip > </network >

表 5-12　路由模式虚拟网络

工具	虚拟网络配置
virt-manager	Mode：Routed，Forward to：Any physical device
Cockpit	不适用

虚拟网络 XML	< network >
	  < name > route1 </name >
	  < forward mode="route"/>
	  < domain name="route1"/>
	  < ip address="192.168.100.1" netmask="255.255.255.0">
	    < dhcp >
	      < range start="192.168.100.128" end="192.168.100.254"/>
	    </dhcp >
	  </ip >
	</network >

表 5-13　SR-IOV 模式虚拟网络

工具	虚拟网络配置
virt-manager	Mode：SR-IOV pool
Cockpit	不适用

虚拟网络 XML	< network >
	  < name > sriov1 </name >
	  < forward mode="hostdev" managed="yes"/>
	</network >

5.4.3　NAT 模式

对于桌面虚拟化或测试环境来讲，NAT 模式是最常用的虚拟网络模式。不需要进行特别配置，此模式的虚拟机就可以访问外部网络，它还允许宿主机与虚拟机之间进行通信。NAT 模式的主要"缺点"是宿主机之外的系统无法访问虚拟机。默认的 default 网络就是 NAT 模式，如图 5-3 所示。

NAT 模式的虚拟网络是在 iptables 的帮助下创建的，其实是使用伪装选项，因此，停止

iptables 会导致虚拟机内部网络的中断。下面通过实验进行验证，代码如下：

```
1 # virsh domiflist centos6.10
  Interface   Type      Source    Model   MAC
  -------------------------------------------------------
  vnet0       network   default   virtio  52:54:00:32:f8:e2

2 # virsh domifaddr centos6.10
  Name     MAC address         Protocol    Address
  -------------------------------------------------------
  vnet0    52:54:00:32:f8:e2   ipv4        192.168.122.18/24

3 # ssh 192.168.122.18 "ping -c 1 www.baidu.com"
  root@192.168.122.18's password：输入虚拟机操作系统 root 的密码
  PING www.wshifen.com (103.235.46.39) 56(84) Bytes of data.
  64 Bytes from 103.235.46.39: icmp_seq=1 ttl=127 time=418 ms

  --- www.wshifen.com ping statistics ---
  1 packets transmitted, 1 received, 0% packet loss, time 418ms
  rtt min/avg/max/mdev = 418.056/418.056/418.056/0.000 ms
```

第 1 行命令的输出结果显示虚拟机 centos6.10 有一个连接到 default 网络的网络接口。

第 2 行命令的输出结果显示这个接口的 IP 地址是 192.168.122.18/24。

第 3 行命令的输出结果验证了两个结论，一个是在宿主机可以访问处于 NAT 网络的虚拟机，另外一个是从这台虚拟机可以访问宿主机外部的网络。

接下来执行的命令如下：

```
4 # systemctl stop firewalld.service

5 # iptables -L -t nat
  Chain PREROUTING (policy ACCEPT)
  target     prot opt source              destination

  Chain INPUT (policy ACCEPT)
  target     prot opt source              destination

  Chain POSTROUTING (policy ACCEPT)
  target     prot opt source              destination

  Chain OUTPUT (policy ACCEPT)
  target     prot opt source              destination

6 # ssh 192.168.122.18 "ping -c 1 www.baidu.com"
  root@192.168.122.18's password:
```

```
       PING www.wshifen.com (103.235.46.39) 56(84) Bytes of data.

       --- www.wshifen.com ping statistics ---
       1 packets transmitted, 0 received, 100 % packet loss, time 10000ms
```

第 4 行命令停止了防火墙服务。从第 5 行命令的输出中可以看出 iptables 的 NAT 表中的所有链的规则均为空。

从第 6 行命令的输出结果可以看出，这台虚拟机现在无法访问宿主机外部的网络。

提示：域名解析和 DHCP 服务是由 dnsmasq 进程提供的，停止防火墙服务不会影响此进程，所以第 6 行命令的输出显示域名解析服务工作正常。

5.4.4 桥接模式

libvirt 不能直接管理桥接模式的网络，所以需要先使用 NetworkManager、Cockpit 等工具在操作系统中创建一个网桥，然后将一个物理网卡（或捆绑在一起的多个物理网卡）分配给网桥作为子接口，最后将虚拟机连接到此网桥（虚拟交换机），如图 5-16 所示。

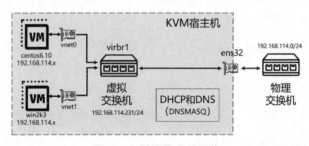

图 5-16　桥接模式的网络

网桥是在 OSI 网络模型的第 2 层上运行的。网桥（虚拟交换机）virbr1 通过 ens32 与宿主机外部的网络相连接。虚拟机 centos6.10、win2k3 与宿主机在同一个子网中，外部物理网络上的主机可以检测到它们并对其进行访问。

5.4.5 隔离模式

顾名思义，这是一种封闭的网络，连接在此虚拟交换机的虚拟机可以彼此通信，也可以与宿主机进行通信，但是它们的流量不会通过宿主机扩散到外部，当然也无法从宿主机外部访问它们，如图 5-17 所示。

5.4.6 路由模式

在路由模式下，宿主机充当路由器并配置路由规则，虚拟机通过虚拟交换机与物理网络相连，如图 5-18 所示。使用此模式有一个关键点：必须在外部的路由器或网关设备上设置正确的 IP 路由，以便答复数据包返回宿主机，例如：需要有"将目标地址是 192.168.100.0/24 的数据包转发给 ens32 的 IP 地址"的路由，否则这些回复数据包将永远不会到达宿主机。

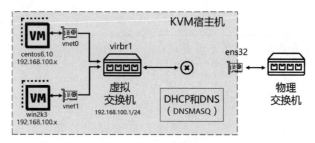

图 5-17　隔离模式的网络

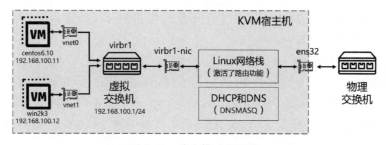

图 5-18　路由模式的网络

提示：除非有特殊的需求（如防火墙的 DMZ 区域），才会创建这种复杂性的网络，否则通常不会使用此模式。

5.4.7　开放模式

开放模式与路由模式类似，区别在于 libvirt 是否为虚拟网络设置防火墙规则。

在路由模式中，libvirt 会配置防火墙规则，例如允许 DHCP、DNS 等流量，而在配置开放模式时，libvirt 不会在此虚拟网络生成任何 iptables 规则，这样就需要用户自行配置。这就是"开放"所代表的含义。

5.4.8　直接附加模式

虚拟机的直接附加模式既没有使用 libvirt 管理的虚拟网络，也没有使用宿主机上的网桥，而是通过宿主机上的 macvtap 驱动程序，将虚拟机的网卡直接附加到宿主机上的指定物理网卡，如图 5-19 所示。

这种模式最大的优点是在物理交换机上可以同时知道宿主机和虚拟机的网卡，从而可以进行统一

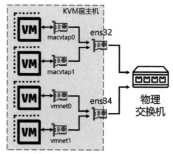

图 5-19　直接附加模式

管理。

直接附加有 4 种模式：VEPA、bridge、private、passthrough，它们在流量控制上有一些差异，默认为 VEPA 模式。

注意：不要将此模式与宿主机设备直通（passthrough）混淆。

5.4.9　PCI 直通与 SR-IOV

KVM Hypervisor 可以将宿主机上符合条件的 PCI 设备分配给虚拟机，这就是 PCI 直通（PCI passthrough）。它可以使虚拟机独占式地访问 PCI 设备，而且不需要或很少需要 KVM 的参与，所以可以提高性能。

目前大多数网卡支持这种特性，通过 PCI 直通将它们分配给虚拟机，从而满足对网络性能要求比较高的应用需求。

但是在服务器上安装 PCI 或 PCI-E 设备的数量总是有限的，而且随着设备数量的增加，也会增加成本。SR-IOV（Single Root Input/Output Virtualization）就是针对这种问题的一种解决方案。它可以将单个物理 PCI 设备划分为多个虚拟的 PCI 设备，然后把它们分配给虚拟机。例如，将宿主机上支持 SR-IOV 功能的网卡划分成多个独立的虚拟网卡，将每个虚拟网卡分配给一个虚拟机使用。

下面就以 Intel 的 X540-AT2 以太网卡为例进行说明。

首先，保证宿主机上有正确的驱动程序，可以识别到这块网卡，然后使用 virt-manager 为一台虚拟机添加硬件，单击 PCI Host Device，右边就会显示一系列可以分配给虚拟机的 PCI 设备。选中所需的 Intel 的 X540 的 Virtual Function，然后单击 Finish 按钮，如图 5-20 所示。这样虚拟机就可以使用这个 PCI 直通设备了。

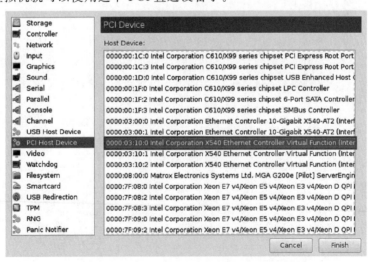

图 5-20　给虚拟机添加支持 SR-IOV 的网卡

5.5 创建和管理隔离的网络

隔离的网络是最简单的网络。本节我们将参考图 5-17 所示的拓扑,通过 virt-manager、Cockpit 和 virsh 来创建和管理一个名为 isolated1 的隔离网络。

5.5.1 通过 virt-manager 创建和管理隔离网络

首先,我们通过 virt-manager 进行创建,然后通过 virsh 及其他操作系统命令进行查看。

(1) 在 virt-manager 的 Edit 菜单上,选择 Connection Details。

(2) 在打开的 Connection Details 菜单中单击 Virtual Networks 选项卡。

(3) 窗口的左侧列出了所有可用的虚拟网络。单击左下角的 + 按钮,会出现 Create a new virtual network 窗口,如图 5-21 所示。

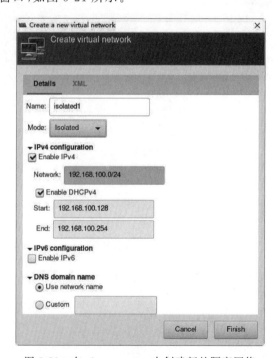

图 5-21 在 virt-manager 中创建新的隔离网络

(4) 设置新虚拟网络的名称,将 Mode 设置为 Isolated,设置适合的 IPv4 configuration。

(5) 单击 XML 标签,会看到虚拟网络的 XML 定义,如图 5-22 所示。

(6) 单击 Finish 按钮完成新虚拟网络的创建。新的虚拟网络的信息如图 5-23 所示。

图 5-22 新的隔离网络的 XML 定义

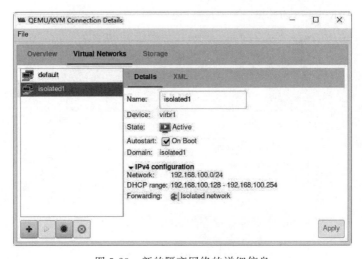

图 5-23 新的隔离网络的详细信息

下面通过 virsh 及其他操作系统命令来查看这个新的虚拟网络,命令如下:

```
1 # virsh net-list
  Name          State      Autostart    Persistent
  -------------------------------------------------
  default       active     yes          yes
  isolated1     active     yes          yes

2 # virsh net-dumpxml isolated1
  <network>
   <name>isolated1</name>
   <uuid>44505edf-d08b-4552-a9f9-0254b80db693</uuid>
   <bridge name='virbr1' stp='on' delay='0'/>
```

```
    <mac address = '52:54:00:9a:14:dd'/>
    <domain name = 'isolated1'/>
    <ip address = '192.168.100.1' netmask = '255.255.255.0'>
      <dhcp>
        <range start = '192.168.100.128' end = '192.168.100.254'/>
      </dhcp>
    </ip>
</network>
```

第 1 行命令的输出显示系统新增加了一个名为 isolated1 的虚拟网络,它是永久的、自动启动的,而且当前状态是已激活的。

第 2 行命令的输出显示了 isolated1 的详细信息,包括以下信息。

(1) name:提供了虚拟网络的名称。

(2) uuid:为虚拟网络提供了全局唯一的标识符。

(3) bridge:name 属性定义了将用于构建虚拟网络的网桥设备的名称。虚拟机将连接到该网桥,从而使它们可以相互通信。由于当前没有 forward 元素,所以这是一个隔离网络。以后会学到 mode 等于 nat、route 或 open 的 forward 元素。属性 stp 指定是否支持生成树协议。属性 delay 指定以秒为单位设置延迟值(默认为 0)。

(4) mac:address 属性定义了一个 MAC(硬件)地址。

(5) domain:name 属性定义了虚拟网络的名称。

(6) ip:address 定义了点分十进制格式的 IPv4 地址或标准冒号分隔的十六进制格式的 IPv6 地址,这些地址将在与虚拟网络关联的网桥设备上进行配置。netmask 定义了子网掩码。

(7) dhcp:指定在此虚拟网络上启用 DHCP 服务。

(8) range:start 和 end 属性指定了要提供给 DHCP 客户端的地址池的边界。

接下来执行的命令如下:

```
3 # ls /etc/libvirt/qemu/networks/
autostart  default.xml  isolated1.xml

4 # cat /etc/libvirt/qemu/networks/isolated1.xml
<!--
WARNING: THIS IS AN AUTO-GENERATED FILE. CHANGES TO IT ARE LIKELY TO BE
OVERWRITTEN AND LOST. Changes to this xml configuration should be made using:
  virsh net-edit isolated1
or other application using the libvirt API.
-->

<network>
  <name>isolated1</name>
```

```
        < uuid > 44505edf - d08b - 4552 - a9f9 - 0254b80db693 </uuid >
        < bridge name = 'virbr1' stp = 'on' delay = '0'/>
        < mac address = '52:54:00:9a:14:dd'/>
        < domain name = 'isolated1'/>
        < ip address = '192.168.100.1' netmask = '255.255.255.0'>
            < dhcp >
                < range start = '192.168.100.128' end = '192.168.100.254'/>
            </dhcp >
        </ip >
    </network >

5 # ls /etc/libvirt/qemu/networks/autostart/ -l
    total 0
    lrwxrwxrwx. 1 root root 14 Jul 30 16:30 default.xml -> ../default.xml
    lrwxrwxrwx. 1 root root 40 Nov 23 17:50 isolated1.xml -> /etc/libvirt/qemu/networks/
    isolated1.xml
```

第 3 行命令的输出显示在 /etc/libvirt/qemu/networks/ 目录中，新增了 1 个 XML 文件 isolated1.xml。不建议直接编辑这个文件，而要使用 virsh net-edit 进行编辑，这是因为它在保存文件时会进行一些必要的检查。

第 5 行命令的输出显示在 /etc/libvirt/qemu/networks/autostart/ 目录中，有一个指定配置文件的符号链接，这样可以保证在启动 libvirtd 守护程序时自动启动这个虚拟网络。

接下来执行的命令如下：

```
6 # ip address
    ...
    8: virbr1: < NO - CARRIER, BROADCAST, MULTICAST, UP > mtu 1500 qdisc noqueue state DOWN group
default qlen 1000
        link/ether 52:54:00:9a:14:dd brd ff:ff:ff:ff:ff:ff
        inet 192.168.100.1/24 brd 192.168.100.255 scope global virbr1
            valid_lft forever preferred_lft forever
    9: virbr1 - nic: < BROADCAST, MULTICAST > mtu 1500 qdisc fq_codel master virbr1 state DOWN
group default qlen 1000
        link/ether 52:54:00:9a:14:dd brd ff:ff:ff:ff:ff:ff

7 # nmcli connection
    NAME    UUID                                            TYPE        DEVICE
    ens32   152beb06 - 47c5 - c5e8 - 95a9 - 385590654382    ethernet    ens32
    virbr0  2ffc46c3 - 9abb - 4071 - b5f8 - 61557313d623    bridge      virbr0
    virbr1  44d4f380 - 9fde - 4d84 - a8c9 - 1a24d159e30b    bridge      virbr1

8 # nmcli device
    DEVICE  TYPE        STATE       CONNECTION
    ens32   ethernet    connected   ens32
```

```
virbr0       bridge     connected    virbr0
virbr1       bridge     connected    virbr1
lo           loopback   unmanaged    --
virbr0-nic   tun        unmanaged    --
virbr1-nic   tun        unmanaged    --
```

第 6 行命令的输出显示系统新增加了 1 个名为 virbr1 的网桥,它的 IP 地址是我们指定的 192.168.100.1/24。此网桥有一个名为 virbr1-nic 的子接口,virbr1 与 virbr1-nic 的 MAC 地址都是 52:54:00:9a:14:dd。

在第 7 行、第 8 行的 nmcli 的子命令的输出中,也会看到 virbr1 与 virbr1-nic。

```
9# route -n
   Kernel IP routing table
   Destination    Gateway          Genmask         Flags  Metric  Ref  Use  Iface
   0.0.0.0        192.168.114.2    0.0.0.0         UG     102     0    0    ens32
   192.168.100.0  0.0.0.0          255.255.255.0   U      0       0    0    virbr1
   192.168.114.0  0.0.0.0          255.255.255.0   U      102     0    0    ens32
   192.168.122.0  0.0.0.0          255.255.255.0   U      0       0    0    virbr0
```

第 9 行命令的输出显示在宿主机的路由表中,新增了一个到 192.168.100.0/24 的路由,接口是 virbr1 交换机,所以可以在宿主机上访问连接到 isolated1 网络的虚拟机。

提示:在创建隔离模式的虚拟网络时,不会引起 iptables 规则的变化。

当不再需要此虚拟网络的时候,可以在 virt-manager 中先单击 ⊙ 按钮停止网络,然后单击 ⊗ 按钮删除网络。

5.5.2 通过 Cockpit 创建和管理隔离网络

通过 Cockpit 创建和管理隔离网络与 virt-manager 很类似。

(1) 单击左边导航中的 Virtual Machines 链接,再单击 Networks 链接。

(2) 单击 Create Virtual Network 链接,然后为新虚拟网络设置属性,然后单击 Create 按钮以完成创建,如图 5-24 所示。需要注意的是:IPv4 Network 选项设置的是新创建网桥的 IP 地址,而不是网段的 IP 地址。

(3) 与 virt-manager 不同,还需要单击 Activate 按钮以激活新创建的虚拟网络。还可以根据需要选中 Run when host boots 检查框,将虚拟网络设置为自动启动,如图 5-25 所示。

当不再需要此虚拟网络的时候,可以单击 Deactive 按钮停止网络,然后单击 Delete 按钮删除网络。

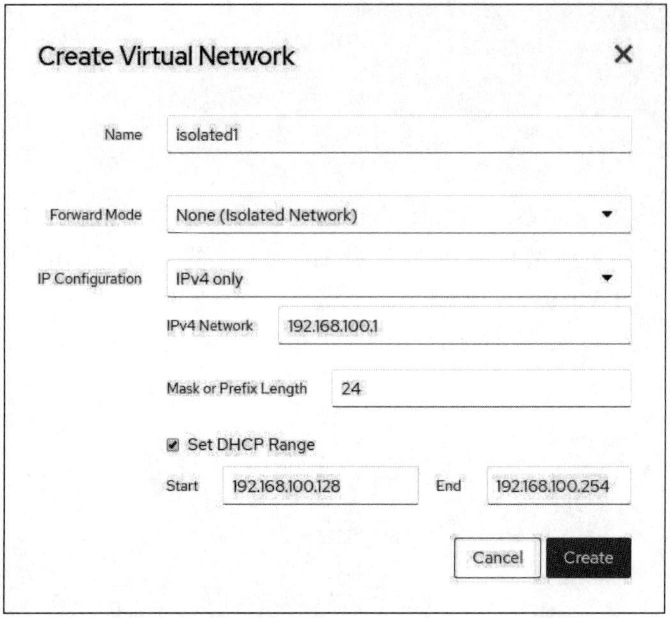

图 5-24 在 Cockpit 中创建新的隔离网络

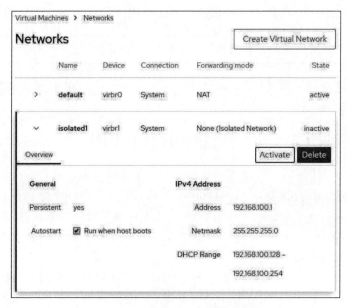

图 5-25 新的隔离网络的详细信息

5.5.3 通过 virsh 创建和管理隔离网络

virsh 管理虚拟网络的子命令多数是以 net 开头的,可以通过 virsh help network 获得这些子命令的清单，通过 virsh help 获得某个子命令的详细帮助，命令如下：

```
1 # virsh help network
  Networking (help keyword 'network'):
    net-autostart      autostart a network
    net-create         create a network from an XML file
    net-define         define an inactive persistent virtual network or modify an existing persistent one from an XML file
    net-destroy        destroy (stop) a network
    net-dhcp-leases    print lease info for a given network
    net-dumpxml        network information in XML
    net-edit           edit XML configuration for a network
    net-event          Network Events
    net-info           network information
    net-list           list networks
    net-name           convert a network UUID to network name
    net-start          start a (previously defined) inactive network
    net-undefine       undefine a persistent network
    net-update         update parts of an existing network's configuration
    net-uuid           convert a network name to network UUID

2 # virsh help net-create
  NAME
    net-create - create a network from an XML file
  SYNOPSIS
    net-create <file>
  DESCRIPTION
    Create a network.
  OPTIONS
    [--file] <string> file containing an XML network description
```

virsh 的 net-create 子命令可以根据 1 个 XML 文件中的设置来创建临时性的（transient）虚拟网络。net-define 子命令既可以创建新的虚拟网络，也可以修改现有的虚拟网络。net-define 子命令创建的新的虚拟网络是持久的，但默认为未激活，需要使用 net-start 子命令进行激活。

下面，我们使用 net-define 子命令来做一个实验，命令如下：

```
3 # vi isolated1.xml
  # 新虚拟网络的定义如下
  <network>
    <name>isolated1</name>
    <domain name="isolated1"/>
    <ip address="192.168.100.1" netmask="255.255.255.0">
      <dhcp>
        <range start="192.168.100.128" end="192.168.100.254"/>
      </dhcp>
```

```
        </ip>
    </network>

4 # virsh net-define isolated1.xml
  Network isolated1 defined from isolated1.xml

5 # virsh net-dumpxml isolated1
  <network>
     <name>isolated1</name>
     <uuid>744c4ba9-cb35-4acc-b577-00b75d4eee52</uuid>
     <bridge name='virbr1' stp='on' delay='0'/>
     <mac address='52:54:00:28:05:b0'/>
     <domain name='isolated1'/>
     <ip address='192.168.100.1' netmask='255.255.255.0'>
        <dhcp>
           <range start='192.168.100.128' end='192.168.100.254'/>
        </dhcp>
     </ip>
  </network>
```

第3行命令在当前目录下创建了一个新文件,在其中为新的虚拟网络设置了一些必要的属性。如果不指定模式,则默认为隔离模式。如果不指定 bridge 属性,则 libvirt 会根据默认值生成一个名称为 virbr 开头的网桥。

第4行命令定义了一个新的网桥。从第5行命令的输出中可以看到 libvirt 会生成一些属性值,包括 uuid、bridge、mac 等。

接下来执行的命令如下:

```
6 # virsh net-list
  Name         State        Autostart    Persistent
  ----------------------------------------------------
  default      active       yes          yes

7 # virsh net-list --all
  Name         State        Autostart    Persistent
  ----------------------------------------------------
  default      active       yes          yes
  isolated1    inactive     no           yes

8 # virsh net-start isolated1
  Network isolated1 started

9 # virsh net-autostart isolated1
  Network isolated1 marked as autostarted
```

```
10 #virsh net-list --all
 Name         State     Autostart     Persistent
 ----------------------------------------------------
 default      active    yes           yes
 isolated1    active    yes           yes
```

由于新增加的虚拟网络并没有被激活,所以它不会出现在第 6 行命令的输出中。第 7 行命令有--all 选项,所以可以看到这个新网络。

第 8 行命令启动这个网络。第 9 行命令将其设置为自动启动。

如果需要修改虚拟网络,则既可以通过 net-edit 子命令来编辑 XML 文件,也可以通过 net-update 子命令根据另外一个 XML 的文件进行更新。

当不再需要某个虚拟网络的时候,可以先使用 net-destroy 子命令停止它,然后使用 net-undefine 子命令删除它的定义。

5.5.4　使用隔离网络

隔离模式允许虚拟机之间相互通信,但是它们无法与外部物理网络进行通信。下面我们通过实验进行验证。

我们将虚拟机 centos6.10 和 win2k3 都更改到 isolated1 这个虚拟网络中。更改虚拟机网络连接的方法很简单,例如在 virt-manager 中更改虚拟机的网络配置即可,如图 5-26 所示。

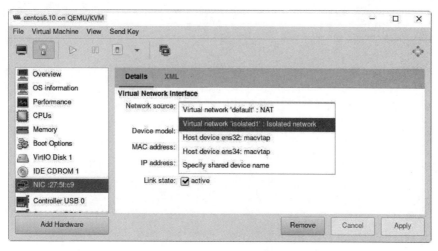

图 5-26　更新虚拟机的网络配置

使用 virsh 的 edit 可以修改虚拟机的 XML 文件。例如将原有虚拟网卡配置进行修改,原配置如下:

```
<interface type = 'network'>
  <mac address = '52:54:00:27:5f:c9'/>
  <source network = 'default'/>
  <model type = 'virtio'/>
  <address type = 'pci' domain = '0x0000' bus = '0x00' slot = '0x03' function = '0x0'/>
</interface>
```

修改后的配置如下:

```
<interface type = 'network'>
  <mac address = '52:54:00:27:5f:c9'/>
  <source network = 'isolated1'/>
  <model type = 'virtio'/>
  <address type = 'pci' domain = '0x0000' bus = '0x00' slot = '0x03' function = '0x0'/>
</interface>
```

最主要是修改<source network='isolated1'/>。修改之后重新启动虚拟机即可生效。如果想修改正在运行的虚拟机网卡配置,则需要使用 update-device 子命令,命令如下:

```
1 # vi new.xml
  <interface type = 'network'>
  <mac address = '52:54:00:32:f8:e2'/>
  <source network = 'isolated1'/>
  <model type = 'virtio'/>
  <address type = 'pci' domain = '0x0000' bus = '0x00' slot = '0x03' function = '0x0'/>
  </interface>

2 # virsh update-device --domain centos6.10 --file new.xml --persistent --live
  Device updated successfully

  # virsh edit centos6.10

3 # virsh domiflist centos6.10
  Interface    Type       Source     Model    MAC
  -------------------------------------------------------------
  vnet0        network    isolated1  virtio   52:54:00:32:f8:e2
```

第 1 行命令创建了包括新配置的 XML 文件。与原有 XML 文件的区别主要是<source network='isolated1'/>这一行。

第 2 行命令使用 XML 文件来更新设备。通过--domain 指定虚拟机的名称、ID 或 uuid,--file 后可指定 XML 文件,--persistent 使变化持久,--live 用于修改正在运行的虚拟机。

从第 3 行命令的输出可以看出虚拟机的网卡切换到 isolated1 虚拟网络了。

提示：如果虚拟机 IP 是动态 IP 地址，则需要等到 DHCP 租期过期后才会更新。如果需要立即更新，就需要在虚拟机操作系统中进行 DHCP 刷新操作了。

思考：一台宿主机上可以拥有多个隔离的网络，属于不同隔离网络的虚拟机是否可能通信呢？如果不可以，则是否可以通过调整配置来使它们之间相互通信呢？

5.6 创建和管理 NAT 的网络

5.6.1 使用多种方式创建 NAT 网络

在 RHEL/CentOS 8 上安装 KVM 虚拟化组件的时候，会自动创建一个名为 default 的 NAT 网络。一台宿主机上可以有多个采用 NAT 模式的虚拟网络，下面我们再创建一个名为 nat2 的 NAT 模式的虚拟网络。

如果使用 virt-manager 来创建，则需要将 Mode 设置为 NAT，可以保持 Forward to 为 Any physical device，设置适合的 IPv4 configuration，如图 5-27 所示。

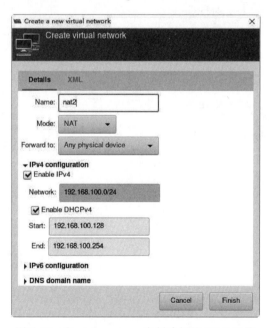

图 5-27　在 virt-manager 中创建新的 NAT 网络

单击 XML 选项卡可以查看新网络的 XML 定义，单击 Finish 按钮完成定义，如图 5-28 所示。

在 Cockpit 创建新的 NAT 网络的方式与 virt-manager 类似，不过需要注意的是：IPv4 Network 选项设置的是新创建网桥的 IP 地址，而不是网段的 IP 地址，如图 5-29 所示。

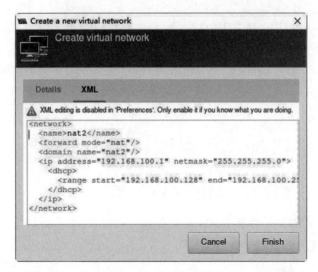

图 5-28 在 virt-manager 查看新的网络的 XML 定义

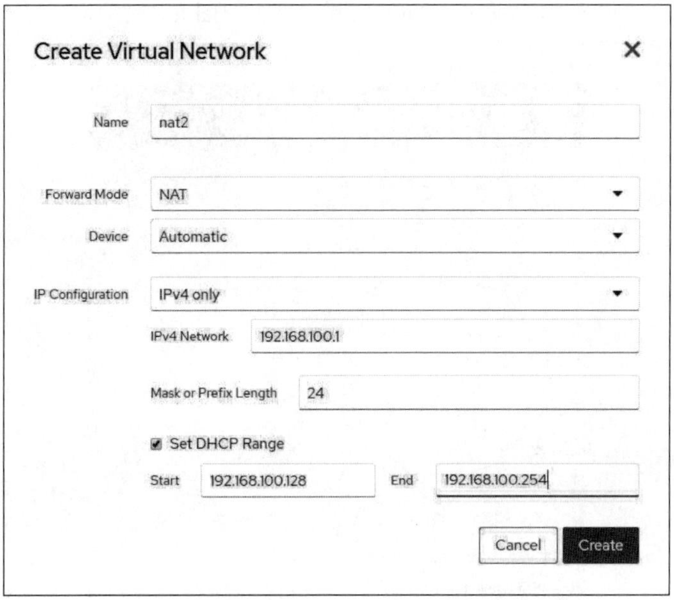

图 5-29 在 Cockpit 中创建新的 NAT 网络

通过 virsh 来创建 NAT 模式的网络，仍然需要使用 XML 格式的配置文件，命令如下：

```
1 # vi new.xml
  <network>
    <name>nat2</name>
    <forward mode="nat"/>
```

```
      <domain name = "nat2"/>
      <ip address = "192.168.100.1" netmask = "255.255.255.0">
        <dhcp>
          <range start = "192.168.100.128" end = "192.168.100.254"/>
        </dhcp>
      </ip>
</network>

2 # virsh net-define new.xml
  Network nat2 defined from new.xml

3 # virsh net-autostart nat2
  Network nat2 marked as autostarted

4 # virsh net-start nat2
  Network nat2 started

5 # virsh net-list --all
   Name       State     Autostart    Persistent
  ----------------------------------------------------------
   default    active    yes          yes
   nat2       active    yes          yes
```

第 1 行命令在当前目录下创建了一个新文件,在其中为新的虚拟网络设置了一些必要的属性。"<forward mode="nat"/>"定义的虚拟网络是 NAT。如果不指定 bridge 属性,则 libvirt 会生成一个以 virbr 开头的网桥。

第 2 行命令定义了一个新的网桥。

第 3 行命令设置为自动启动,第 4 行命令用于启动这个网络。

接下来执行的命令如下:

```
6 # ip address
  …
  6: virbr1: <NO-CARRIER,BROADCAST,MULTICAST,UP> mtu 1500 qdisc noqueue state DOWN group default qlen 1000
    link/ether 52:54:00:4d:94:67 brd ff:ff:ff:ff:ff:ff
    inet 192.168.100.1/24 brd 192.168.100.255 scope global virbr1
    valid_lft forever preferred_lft forever
  7: virbr1-nic: <BROADCAST,MULTICAST> mtu 1500 qdisc fq_codel master virbr1 state DOWN group default qlen 1000
    link/ether 52:54:00:4d:94:67 brd ff:ff:ff:ff:ff:ff

7 # virsh net-dumpxml nat2
  <network>
    <name>nat2</name>
```

```xml
    <uuid>8a54c7f3-e910-4d12-8312-34a11467ac86</uuid>
    <forward mode='nat'>
      <nat>
        <port start='1024' end='65535'/>
      </nat>
    </forward>
    <bridge name='virbr1' stp='on' delay='0'/>
    <mac address='52:54:00:4d:94:67'/>
    <domain name='nat2'/>
    <ip address='192.168.100.1' netmask='255.255.255.0'>
      <dhcp>
        <range start='192.168.100.128' end='192.168.100.254'/>
      </dhcp>
    </ip>
</network>
```

第 6 行命令的输出显示：新增 1 个名为 virbr1 的网桥，它有 1 个名为 virbr1-nic 的子接口。网桥 virbr1 与 virbr1-nic 的 MAC 地址相同。

第 7 行命令会输出新网桥的详细信息，会看到有些属性使用了默认值或是自动生成的。

接下来执行的命令如下：

```
8 # iptables -L -t nat -n
  Chain PREROUTING (policy ACCEPT)
  target     prot opt source              destination

  Chain INPUT (policy ACCEPT)
  target     prot opt source              destination

  Chain POSTROUTING (policy ACCEPT)
  target     prot opt source              destination
  RETURN     all  --  192.168.100.0/24    224.0.0.0/24
  RETURN     all  --  192.168.100.0/24    255.255.255.255
  MASQUERADE tcp  --  192.168.100.0/24    !192.168.100.0/24    masq ports: 1024-65535
  MASQUERADE udp  --  192.168.100.0/24    !192.168.100.0/24    masq ports: 1024-65535
  MASQUERADE all  --  192.168.100.0/24    !192.168.100.0/24
  RETURN     all  --  192.168.122.0/24    224.0.0.0/24
  RETURN     all  --  192.168.122.0/24    255.255.255.255
  MASQUERADE tcp  --  192.168.122.0/24    !192.168.122.0/24    masq ports: 1024-65535
  MASQUERADE udp  --  192.168.122.0/24    !192.168.122.0/24    masq ports: 1024-65535
  MASQUERADE all  --  192.168.122.0/24    !192.168.122.0/24

  Chain OUTPUT (policy ACCEPT)
  target     prot opt source              destination
```

```
9 # route -n
Kernel IP routing table
Destination      Gateway         Genmask         Flags  Metric  Ref  Use  Iface
0.0.0.0          192.168.114.2   0.0.0.0         UG     100     0    0    ens32
192.168.100.0    0.0.0.0         255.255.255.0   U      0       0    0    virbr1
192.168.114.0    0.0.0.0         255.255.255.0   U      100     0    0    ens32
192.168.122.0    0.0.0.0         255.255.255.0   U      0       0    0    virbr0
```

libvirt 会根据 NAT 模式的虚拟网络的配置来修改 iptables 的 NAT 规则和系统的路由。

从第 8 行命令的输出可以看到在 iptables 的 NAT 表的 POSTROUTING 链中有多条针对 192.168.100.0/24 的规则。正是由于这些规则才使连接到 nat2 网络中的虚拟机可以访问外部的网络。

第 9 行命令的输出显示在宿主机的路由表中，新增了一个到 192.168.100.0/24 的路由，目标地址是这个网络的数据包将通过 virbr1 交换机输送出去。这样，就可以在宿主机上访问连接到 nat2 网络的虚拟机了，而且属于 default 网络的虚拟机也可以访问 nat2 网络的虚拟机。

5.6.2 使用 NAT 网络

在实验环境中，准备使用两台虚拟机来做实验。win2k3 还是属于 default 网络，将 centos6.10 修改为 nat2，命令如下：

```
1 # virsh edit centos6.10
  …
  <interface type = 'network'>
    <mac address = '52:54:00:32:f8:e2'/>
    <source network = 'nat2'/>
    <model type = 'virtio'/>
    <address type = 'pci' domain = '0x0000' bus = '0x00' slot = '0x03' function = '0x0'/>
  </interface>
  …

2 # virsh start centos6.10

3 # virsh start win2k3
```

第 1 行命令用于编辑虚拟配置的 XML 文件，修改网卡的属性 source，并将其修改为 <source network = 'nat2'/>。

第 2 行、第 3 行命令分别启动了这 2 台属于不同虚拟网络的虚拟机。

接下来执行的命令如下：

```
4 # ip address
  ...
  8: vnet0: < BROADCAST,MULTICAST,UP,LOWER_UP > mtu 1500 qdisc fq_codel master virbr1 state UNKNOWN group default qlen 1000
    link/ether fe:54:00:27:5f:c9 brd ff:ff:ff:ff:ff:ff
    inet6 fe80::fc54:ff:fe27:5fc9/64 scope link
    valid_lft forever preferred_lft forever
  9: vnet1: < BROADCAST,MULTICAST,UP,LOWER_UP > mtu 1500 qdisc fq_codel master virbr0 state UNKNOWN group default qlen 1000
    link/ether fe:54:00:9d:57:9c brd ff:ff:ff:ff:ff:ff
    inet6 fe80::fc54:ff:fe9d:579c/64 scope link
    valid_lft forever preferred_lft forever
  ...

5 # virsh domiflist centos6.10
 Interface    Type       Source     Model     MAC
 -------------------------------------------------------------
 vnet0        network    nat2       virtio    52:54:00:27:5f:c9

6 # virsh domifaddr centos6.10
 Name     MAC address           Protocol     Address
 -------------------------------------------------------------
 vnet0    52:54:00:27:5f:c9     ipv4         192.168.100.139/24

7 # virsh domiflist win2k3
 Interface    Type       Source     Model     MAC
 -------------------------------------------------------------
 vnet1        network    default    e1000     52:54:00:9d:57:9c

8 # virsh domifaddr win2k3
 Name     MAC address           Protocol     Address
 -------------------------------------------------------------
 vnet1    52:54:00:9d:57:9c     ipv4         192.168.122.41/24
```

从第 3 行命令的输出可以看到宿主机上新增了 2 个 TAP 类型的虚拟网络接口 vnet0 和 vnet1，它们分别是网桥 virbr1 和 virbr0 的子接口。

从第 5 行至第 8 行的输出可以看出：虚拟机 centos6.10 连接到虚拟网络 nat2，它的 IP 地址是 DNSMASQ 分配的 192.168.100.139/2。虚拟机 win2k3 连接到虚拟网络 default，它的 IP 地址是 DNSMASQ 分配的 192.168.122.41/24。

提示：宿主机上名称为 vnet 开头的 TAP 设备与虚拟机的以太网卡是"一对"设备，它们的 MAC 地址后面 5 组 2 位十六进制数字是相同的，例如：vnet0 与 centos6.10 的 eth0 的 MAC 地址都是以 "54:00:27:5f:c9" 结束。

下面,我们连接到虚拟机 centos6.10(192.168.100.139)来做一些测试,命令如下:

```
 9 # ssh 192.168.100.139 "ping -c 2 www.baidu.com"
   root@192.168.100.139's password:
   PING www.wshifen.com (103.235.46.39) 56(84) Bytes of data.
   64 Bytes from 103.235.46.39: icmp_seq = 1 ttl = 127 time = 278 ms
   64 Bytes from 103.235.46.39: icmp_seq = 2 ttl = 127 time = 277 ms

   --- www.wshifen.com ping statistics ---
   2 packets transmitted, 2 received, 0% packet loss, time 1281ms
   rtt min/avg/max/mdev = 277.802/278.094/278.387/0.603 ms
10 # ssh 192.168.100.139 "ping -c 2 192.168.122.41"
   root@192.168.100.139's password:
   PING 192.168.122.41 (192.168.122.41) 56(84) Bytes of data.
   64 Bytes from 192.168.122.41: icmp_seq = 1 ttl = 127 time = 1.96 ms
   64 Bytes from 192.168.122.41: icmp_seq = 2 ttl = 127 time = 2.10 ms

   --- 192.168.122.41 ping statistics ---
   2 packets transmitted, 2 received, 0% packet loss, time 1005ms
   rtt min/avg/max/mdev = 1.962/2.034/2.107/0.085 ms
```

从第 9 行命令的输出中可以看出:从宿主机可以访问连接到虚拟网络 nat2 的虚拟机,而这些虚拟机可以通过宿主机的 NAT 功能访问外部网络资源。

从第 10 行命令的输出可以看出:连接到虚拟网络 nat2 的虚拟机也可以访问同一宿主机上的其他 NAT 网络中的虚拟机(default 网络中的 win2k3)。

5.7 创建和管理桥接的网络

桥接模式的网络是很常见的,外部网络上的主机可以访问这种连接模式网桥中的虚拟机。

下面按图 5-16 所示的拓扑来做一个桥接模式网络的实验。

5.7.1 在宿主机上创建网桥

执行的命令如下:

```
1 # ip address
  1: lo: <LOOPBACK,UP,LOWER_UP> mtu 65536 qdisc noqueue state UNKNOWN group default qlen 1000
     link/loopback 00:00:00:00:00:00 brd 00:00:00:00:00:00
     inet 127.0.0.1/8 scope host lo
        valid_lft forever preferred_lft forever
     inet6 ::1/128 scope host
```

```
            valid_lft forever preferred_lft forever
    2: ens32: <BROADCAST,MULTICAST,UP,LOWER_UP> mtu 1500 qdisc fq_codel state UP group default
qlen 1000
        link/ether 00:0c:29:b3:73:36 brd ff:ff:ff:ff:ff:ff
        inet 192.168.114.231/24 brd 192.168.114.255 scope global noprefixroute ens32
            valid_lft forever preferred_lft forever
        inet6 fe80::c589:de19:7895:d32e/64 scope link noprefixroute
            valid_lft forever preferred_lft forever
    3: virbr0: <NO-CARRIER,BROADCAST,MULTICAST,UP> mtu 1500 qdisc noqueue state DOWN group
default qlen 1000
        link/ether 52:54:00:fd:b2:60 brd ff:ff:ff:ff:ff:ff
        inet 192.168.122.1/24 brd 192.168.122.255 scope global virbr0
            valid_lft forever preferred_lft forever
    4: virbr0-nic: <BROADCAST,MULTICAST> mtu 1500 qdisc fq_codel master virbr0 state DOWN
group default qlen 1000
        link/ether 52:54:00:fd:b2:60 brd ff:ff:ff:ff:ff:ff
```

第 1 行命令会显示创建新网桥之前网络接口的信息。IP 地址 192.168.114.231/24 是与 ens32 相关联的。

使用 Cockpit 来创建网桥是最简单的方法。新网桥的名称是 virbr1。将宿主机上唯一的物理网卡 en32 作为它的子接口,如图 5-30 所示。

图 5-30　使用 Cockpit 创建新的网桥

接下来执行的命令如下:

```
2 # ip address
    1: lo: <LOOPBACK,UP,LOWER_UP> mtu 65536 qdisc noqueue state UNKNOWN group default qlen 1000
        link/loopback 00:00:00:00:00:00 brd 00:00:00:00:00:00
        inet 127.0.0.1/8 scope host lo
```

```
        valid_lft forever preferred_lft forever
    inet6 ::1/128 scope host
        valid_lft forever preferred_lft forever
2: ens32: <BROADCAST,MULTICAST,UP,LOWER_UP> mtu 1500 qdisc fq_codel master virbr1 state UP group default qlen 1000
    link/ether 00:0c:29:b3:73:36 brd ff:ff:ff:ff:ff:ff
3: virbr0: <NO-CARRIER,BROADCAST,MULTICAST,UP> mtu 1500 qdisc noqueue state DOWN group default qlen 1000
    link/ether 52:54:00:fd:b2:60 brd ff:ff:ff:ff:ff:ff
    inet 192.168.122.1/24 brd 192.168.122.255 scope global virbr0
        valid_lft forever preferred_lft forever
4: virbr0-nic: <BROADCAST,MULTICAST> mtu 1500 qdisc fq_codel master virbr0 state DOWN group default qlen 1000
    link/ether 52:54:00:fd:b2:60 brd ff:ff:ff:ff:ff:ff
5: virbr1: <BROADCAST,MULTICAST,UP,LOWER_UP> mtu 1500 qdisc noqueue state UP group default qlen 1000
    link/ether 00:0c:29:b3:73:36 brd ff:ff:ff:ff:ff:ff
    inet 192.168.114.231/24 brd 192.168.114.255 scope global noprefixroute virbr1
        valid_lft forever preferred_lft forever
    inet6 fe80::17b3:2554:31a6:df6e/64 scope link noprefixroute
        valid_lft forever preferred_lft forever

3 # nmcli connection
NAME    UUID                                    TYPE      DEVICE
virbr1  66629af3-7afb-4cbd-9170-020a9f9cb805    bridge    virbr1
virbr0  665697d3-824e-4832-b057-14acf8d379e0    bridge    virbr0
ens32   152beb06-47c5-c5e8-95a9-385590654382    ethernet  ens32
```

第 2 行命令输出了创建新网桥之后网络接口的信息。IP 地址 192.168.114.231/24 与 virbr1 相关联，ens32 则成为 virbr1 的子接口。网桥 virbr1 与第 1 个子接口也就是 ens32 的 MAC 地址是相同的，都是"00:0c:29:b3:73:36"。

注意：如果宿主机上仅有一个网卡，而且在远程通过 nmcli 或网络接口配置文件来创建网桥，就要特别小心，因为错误的配置会导致无法再进行远程连接。

接下来执行的命令如下：

```
4 # cat /etc/sysconfig/network-scripts/ifcfg-virbr1
    STP=no
    TYPE=Bridge
    PROXY_METHOD=none
    BROWSER_ONLY=no
    BOOTPROTO=none
    IPADDR=192.168.114.231
```

```
        PREFIX = 24
        GATEWAY = 192.168.114.2
        DNS1 = 8.8.8.8
        DEFROUTE = yes
        IPV4_FAILURE_FATAL = no
        IPV6INIT = yes
        IPV6_AUTOCONF = yes
        IPV6_DEFROUTE = yes
        IPV6_FAILURE_FATAL = no
        IPV6_ADDR_GEN_MODE = stable-privacy
        NAME = virbr1
        UUID = 66629af3-7afb-4cbd-9170-020a9f9cb805
        DEVICE = virbr1
        ONBOOT = yes
        AUTOCONNECT_SLAVES = yes

5  # cat /etc/sysconfig/network-scripts/ifcfg-ens32
        TYPE = Ethernet
        PROXY_METHOD = none
        BROWSER_ONLY = no
        DEFROUTE = yes
        IPV4_FAILURE_FATAL = no
        NAME = ens32
        DEVICE = ens32
        ONBOOT = yes
        IPADDR = 192.168.114.231
        PREFIX = 24
        GATEWAY = 192.168.114.2
        DNS1 = 8.8.8.8
        UUID = 152beb06-47c5-c5e8-95a9-385590654382
        BRIDGE = virbr1
```

第 4 行命令会显示新网桥 virbr1 的网络接口配置文件的内容，其中最重要的属性是 TYPE=Bridge。

第 5 行命令显示了配置文件 ifcfg-ens32 的变化。这个修改后的配置文件不是特别"干净"，有一些无用信息。不过最关键的属性是 BRIDGE=virbr1。

5.7.2 使用网桥

以虚拟机 centos6.10 为例，我们通过 Cockpit 来修改其网络接口的配置。在 Interface Type 中选择 Bridge to LAN，然后在 Source 中选择 virbr1，单击 Save 按钮保存更改，如图 5-31 所示。

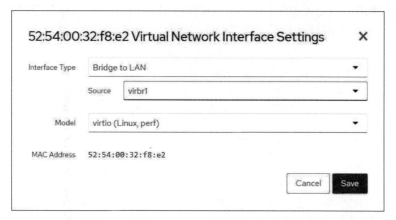

图 5-31　使用 Cockpit 更新虚拟机的网络配置

执行的命令如下：

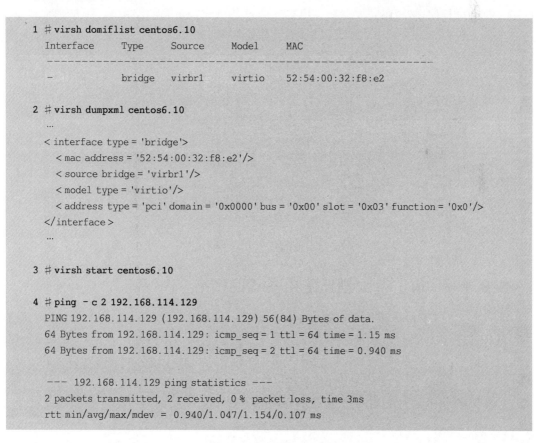

第 1 行命令的输出内容显示了此虚拟机已经连接网桥 virbr1。
第 2 行命令的输出内容显示了它的网桥接口类型是＜interface type＝'bridge'＞。

第 3 行命令用于启动虚拟机。这时，虚拟机 centos6.10 会从宿主机所在物理网络中的 DHCP 服务器获得 IP 地址。我们可以登录到此虚拟机查看其 IP 地址，在本示例中是 192.168.114.129。

第 4 行命令用于对这个 IP 地址进行测试。当然外部物理网络上的其他主机也可以访问此虚拟机。

5.8 创建和管理路由的网络

路由模式不是很常见。在这种模式中，宿主机充当虚拟网络与外部物理网络之间的路由器，而且会自动配置正确的路由条目。这种模式的关键点是在宿主机之外的路由设置，需要用到虚拟网络的"返程"路由。

我们将根据图 5-32 所示的拓扑来做一个实验。在这个实验中，宿主机网卡 ens32 的 IP 地址是 192.168.1.231/24，需要创建的新虚拟网络的 IP 地址是 192.168.100.0/24。除了在宿主机上进行配置外，还需要在宿主机之外的设备上配置正确的路由条目，在此实验中，将在外部主机 192.168.1.48/24 上添加"将目标地址是 192.168.100.0/24 的数据包转发给 192.168.1.231"的路由条目。

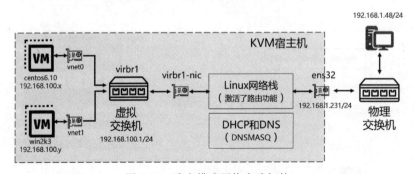

图 5-32 路由模式网络实验拓扑

5.8.1 在宿主机上创建路由模式的网络

当前版本的 Cockpit 虚拟化插件（cockpit-machines-211.3-1.el8.noarch）不支持创建路由模式的网络，但是可以正确地显示路由模式的网络。我们可以通过 virt-manager 来创建路由模式的网络。创建时，将 Mode 设置为 Routed，可以保持 Forward to 为 Any physical device，设置适合的 IPv4 configuration，如图 5-33 所示。

仍然使用 virsh 的 net-define 子命令来创建，创建时需要使用 XML 格式文件的内容如下：

```
<network>
  <name>route1</name>
  <forward mode="route"/>
```

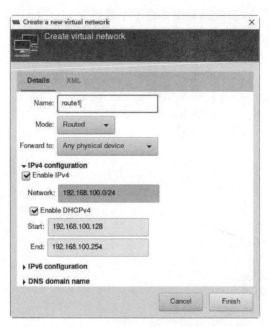

图 5-33　使用 virt-manager 创建路由模式网络

```
    < domain name = "route1"/>
    < ip address = "192.168.100.1" netmask = "255.255.255.0">
      < dhcp >
        < range start = "192.168.100.128" end = "192.168.100.254"/>
      </dhcp>
    </ip>
</network>
```

libvirt 在创建新的路由模式网络的时候，会在宿主机上创建新网桥及其子接口、增加路由中的路由条目、增加防火墙规则，执行的命令如下：

```
1 # ip address
  1: lo: < LOOPBACK,UP,LOWER_UP > mtu 65536 qdisc noqueue state UNKNOWN group default qlen 1000
     link/loopback 00:00:00:00:00:00 brd 00:00:00:00:00:00
     inet 127.0.0.1/8 scope host lo
        valid_lft forever preferred_lft forever
     inet6 ::1/128 scope host
        valid_lft forever preferred_lft forever
  2: ens32: < BROADCAST,MULTICAST,UP,LOWER_UP > mtu 1500 qdisc fq_codel state UP group default qlen 1000
     link/ether 00:0c:29:b3:73:36 brd ff:ff:ff:ff:ff:ff
     inet 192.168.1.231/24 brd 192.168.1.255 scope global noprefixroute ens32
        valid_lft forever preferred_lft forever
```

```
            inet6 fe80::20c:29ff:feb3:7336/64 scope link
               valid_lft forever preferred_lft forever
    3: virbr0: < NO - CARRIER,BROADCAST,MULTICAST,UP > mtu 1500 qdisc noqueue state DOWN group
default qlen 1000
          link/ether 52:54:00:fd:b2:60 brd ff:ff:ff:ff:ff:ff
          inet 192.168.122.1/24 brd 192.168.122.255 scope global virbr0
              valid_lft forever preferred_lft forever
    4: virbr0 - nic: < BROADCAST,MULTICAST > mtu 1500 qdisc fq_codel master virbr0 state DOWN
group default qlen 1000
          link/ether 52:54:00:fd:b2:60 brd ff:ff:ff:ff:ff:ff
    5: virbr1: < NO - CARRIER,BROADCAST,MULTICAST,UP > mtu 1500 qdisc noqueue state DOWN group
default qlen 1000
          link/ether 52:54:00:d2:32:33 brd ff:ff:ff:ff:ff:ff
          inet 192.168.100.1/24 brd 192.168.100.255 scope global virbr1
              valid_lft forever preferred_lft forever
    6: virbr1 - nic: < BROADCAST,MULTICAST > mtu 1500 qdisc fq_codel master virbr1 state DOWN
group default qlen 1000
          link/ether 52:54:00:d2:32:33 brd ff:ff:ff:ff:ff:ff


2 # virsh net - list
   Name          State        Autostart        Persistent
   ----------------------------------------------------------------
   default       active       yes              yes
   route1        active       yes              yes


3 # virsh net - info route1
   Name:           route1
   UUID:           e166c9be - b791 - 4ea4 - 814a - 6f1c99326eea
   Active:         yes
   Persistent:     yes
   Autostart:      yes
   Bridge:         virbr1


4 # virsh net - dumpxml route1
   < network >
     < name > route1 </name >
     < uuid > e166c9be - b791 - 4ea4 - 814a - 6f1c99326eea </uuid >
     < forward mode = 'route'/>
     < bridge name = 'virbr1' stp = 'on' delay = '0'/>
     < mac address = '52:54:00:d2:32:33'/>
     < domain name = 'route1'/>
     < ip address = '192.168.100.1' netmask = '255.255.255.0'>
       < dhcp >
          < range start = '192.168.100.128' end = '192.168.100.254'/>
       </dhcp >
     </ip >
   </network >
```

第 1 行命令的输出内容显示了新网桥的名称 virbr1,其 IP 地址是 192.168.100.1/24。virbr1 有 1 个接口 virbr1-nic。

第 2 行命令的输出内容显示了当前虚拟网络的列表。

第 3 行、第 4 行命令的输出内容显示了新虚拟网络的详细信息。

接下来执行的命令如下:

```
5 # route -n
Kernel IP routing table
Destination     Gateway         Genmask         Flags Metric Ref   Use Iface
0.0.0.0         192.168.1.2     0.0.0.0         UG    100    0     0 ens32
192.168.1.0     0.0.0.0         255.255.255.0   U     100    0     0 ens32
192.168.100.0   0.0.0.0         255.255.255.0   U     0      0     0 virbr1
192.168.122.0   0.0.0.0         255.255.255.0   U     0      0     0 virbr0
```

在第 5 行命令的输出中看到了新增加的路由条目:目标是 192.168.100.0/24 的数据将被发送给 virbr1,也就是新的虚拟网络。

接下来执行的命令如下:

```
6 # iptables -L -n -v
Chain INPUT (policy ACCEPT 202 packets, 17430 Bytes)
 target     prot opt in     out     source          destination
 ACCEPT     udp  --  virbr1 *       0.0.0.0/0       0.0.0.0/0       udp dpt:53
 ACCEPT     tcp  --  virbr1 *       0.0.0.0/0       0.0.0.0/0       tcp dpt:53
 ACCEPT     udp  --  virbr1 *       0.0.0.0/0       0.0.0.0/0       udp dpt:67
 ACCEPT     tcp  --  virbr1 *       0.0.0.0/0       0.0.0.0/0       tcp dpt:67
 ACCEPT     udp  --  virbr0 *       0.0.0.0/0       0.0.0.0/0       udp dpt:53
 ACCEPT     tcp  --  virbr0 *       0.0.0.0/0       0.0.0.0/0       tcp dpt:53
 ACCEPT     udp  --  virbr0 *       0.0.0.0/0       0.0.0.0/0       udp dpt:67
 ACCEPT     tcp  --  virbr0 *       0.0.0.0/0       0.0.0.0/0       tcp dpt:67

Chain FORWARD (policy ACCEPT 0 packets, 0 Bytes)
 target     prot opt in     out            source          destination
 ACCEPT     all  --  *      virbr1         0.0.0.0/0       192.168.100.0/24
 ACCEPT     all  --  virbr1 *              192.168.100.0/24 0.0.0.0/0
 ACCEPT     all  --  virbr1 virbr1         0.0.0.0/0       0.0.0.0/0
 REJECT     all  --  *      virbr1         0.0.0.0/0       0.0.0.0/0       reject-with icmp-port-unreachable
 REJECT     all  --  virbr1 *              0.0.0.0/0       0.0.0.0/0       reject-with icmp-port-unreachable
 ACCEPT     all  --  *      virbr0         0.0.0.0/0       192.168.122.0/24 ctstate RELATED,ESTABLISHED
 ACCEPT     all  --  virbr0 *              192.168.122.0/24 0.0.0.0/0
 ACCEPT     all  --  virbr0 virbr0         0.0.0.0/0       0.0.0.0/0
```

```
            REJECT     all  --  *  virbr0              0.0.0.0/0      0.0.0.0/0      reject-with
icmp-port-unreachable
            REJECT     all  --  virbr0 *               0.0.0.0/0      0.0.0.0/0      reject-with
icmp-port-unreachable
    Chain OUTPUT (policy ACCEPT 177 packets, 21663 Bytes)
        target prot opt in out source destination
        ACCEPT udp  --  *  virbr1 0.0.0.0/0 0.0.0.0/0 udp dpt:68
        ACCEPT udp  --  *  virbr0 0.0.0.0/0 0.0.0.0/0 udp dpt:68
```

从第 6 行命令的输出中可以看到新增加的防火墙规则。

（1）INPUT 链：允许 virbr1 入站的 DNS(53 端口)和 DHCP(67 端口)的包。

（2）FORWARD 链：允许源是 virbr1 出站及相关联的返回包，源与目标都是 virbr1 的包，拒绝 icmp-port-unreachable 类型的包。

（3）OUTPUT 链：允许 virbr1 出站的 DHCP(68 端口)的包。

5.8.2 使用路由模式的网络

将虚拟机 centos6.10 调整到新增加的虚拟网络，命令如下：

```
1  # virsh edit centos6.10
   ...
       <interface type='network'>
         <mac address='52:54:00:32:f8:e2'/>
         <source network='route1'/>
         <model type='virtio'/>
         <address type='pci' domain='0x0000' bus='0x00' slot='0x03' function='0x0'/>
       </interface>
   ...

2  # virsh start centos6.10

3  # virsh domiflist centos6.10
   Interface    Type       Source    Model     MAC
   ------------------------------------------------------------
   vnet0        network    route1    virtio    52:54:00:32:f8:e2

4  # virsh domifaddr centos6.10
   Name         MAC address            Protocol     Address
   ------------------------------------------------------------
   vnet0        52:54:00:32:f8:e2      ipv4         192.168.100.182/24
```

第 1 行命令用于编辑虚拟机的配置文件，其核心属性是 <source network='route1'/>。

第 2 行命令用于启动这台虚拟机。

第 3 行命令的输出显示此虚拟机已经切换到新的名为 route1 的虚拟网络中了。

第 4 行命令显示虚拟机的 IP 地址是 192.168.100.182/24，它是由宿主机的 DNSMASQ 组件分配的。

接下来执行的命令如下：

```
5 # ssh 192.168.100.182 "route -n"
root@192.168.100.182's password:
Kernel IP routing table
Destination     Gateway         Genmask         Flags Metric Ref    Use Iface
192.168.100.0   0.0.0.0         255.255.255.0   U     0      0        0 eth0
169.254.0.0     0.0.0.0         255.255.0.0     U     1002   0        0 eth0
0.0.0.0         192.168.100.1   0.0.0.0         UG    0      0        0 eth0

6 # ssh 192.168.100.182 "ping -c 2 192.168.1.231"
root@192.168.100.182's password:
PING 192.168.1.231 (192.168.1.231) 56(84) Bytes of data.
64 Bytes from 192.168.1.231: icmp_seq=1 ttl=64 time=0.300 ms
64 Bytes from 192.168.1.231: icmp_seq=2 ttl=64 time=0.582 ms

--- 192.168.1.231 ping statistics ---
2 packets transmitted, 2 received, 0% packet loss, time 1002ms
rtt min/avg/max/mdev = 0.300/0.441/0.582/0.141 ms

7 # ssh 192.168.100.182 "ping -c 2 192.168.1.48"
root@192.168.100.182's password:
PING 192.168.1.48 (192.168.1.48) 56(84) Bytes of data.

--- 192.168.1.48 ping statistics ---
2 packets transmitted, 0 received, 100% packet loss, time 11003ms
```

第 5 行命令的输出显示了虚拟机操作系统中的路由表，其缺省网关是 192.168.100.1，也就是宿主中的 virbr1。

第 6 行命令从虚拟机中 ping 宿主机的 ens32 的 IP 地址，由于宿主机知道 192.168.100.0/24 在哪里，所以可以正常返回响应。

第 7 行命令从虚拟机中 ping 宿主机之外的主机(192.168.1.48)，从输出信息中可以看到：没有收到响应的包(0 received)。

这是由于在外部主机 192.168.1.48 上没有到 192.168.100.0/24 的正确路由。

在本实验中，192.168.1.48 是一台 Windows 10 的计算机，所以可以通过 route print 查看其路由表，命令如下：

```
C:\WINDOWS\system32> route print -4
===========================================================================
接口列表
  4...38 d5 47 b9 33 c6 ......Realtek PCIe GbE Family Controller
```

```
   1...........................Software Loopback Interface 1
===========================================================================

IPv4 路由表
===========================================================================
活动路由:
网络目标              网络掩码              网关              接口              跃点数
0.0.0.0              0.0.0.0              192.168.1.1      192.168.1.48     291
127.0.0.0            255.0.0.0            在链路上          127.0.0.1        331
127.0.0.1            255.255.255.255      在链路上          127.0.0.1        331
127.255.255.255      255.255.255.255      在链路上          127.0.0.1        331
192.168.1.0          255.255.255.0        在链路上          192.168.1.48     291
192.168.1.48         255.255.255.255      在链路上          192.168.1.48     291
192.168.1.255        255.255.255.255      在链路上          192.168.1.48     291
224.0.0.0            240.0.0.0            在链路上          127.0.0.1        331
224.0.0.0            240.0.0.0            在链路上          192.168.1.48     291
255.255.255.255      255.255.255.255      在链路上          127.0.0.1        331
255.255.255.255      255.255.255.255      在链路上          192.168.1.48 V 291
===========================================================================
永久路由:
  网络地址            网络掩码             网关地址           跃点数
  0.0.0.0            0.0.0.0              192.168.1.1       默认
===========================================================================
```

由于在路由表中没有与 192.168.100.0/24 相匹配的条目,所以它会将响应 192.168.100.0/24 的包发给默认网关(192.168.1.1),而不是宿主机的 IP 地址 192.168.1.231。在本实验中,可以通过添加静态路由来解决这个问题,命令如下:

```
C:\WINDOWS\system32> route add 192.168.100.0 MASK 255.255.255.0 192.168.1.231

C:\WINDOWS\system32> route print -4 | find "192.168.100"
      192.168.100.0    255.255.255.0    192.168.1.231    192.168.1.48    36
```

这样就可以 ping 通了,命令如下:

```
9 # ssh 192.168.100.182 "ping -c 2 192.168.1.48"
  root@192.168.100.182's password:
  PING 192.168.1.48 (192.168.1.48) 56(84) Bytes of data.
  64 Bytes from 192.168.1.48: icmp_seq=1 ttl=127 time=0.857 ms
  64 Bytes from 192.168.1.48: icmp_seq=2 ttl=127 time=1.86 ms

  --- 192.168.1.48 ping statistics ---
  2 packets transmitted, 2 received, 0% packet loss, time 1004ms
  rtt min/avg/max/mdev = 0.857/1.363/1.869/0.506 ms
```

5.9　创建和管理开放的网络

开放模式与路由模式类似,其主要区别是 libvirt 不会为开放模式设置防火墙规则。下面通过实验来查看这种模式,命令如下:

```
1 # vi new.xml
  < network >
    < name > open1 </name >
    < forward mode = "open"/>
    < domain name = "open1"/>
    < ip address = "192.168.100.1" netmask = "255.255.255.0">
      < dhcp >
        < range start = "192.168.100.128" end = "192.168.100.254"/>
      </dhcp >
    </ip >
  </network >

2 # virsh net - define new.xml

3 # virsh net - autostart open1

4 # virsh net - start open1

5 # virsh net - list
  Name      State     Autostart    Persistent
  -----------------------------------------------------------
  default   active    yes          yes
  open1     active    yes          yes
```

第 1 行命令用于创建新网络的 XML 定义文件。

第 2 行命令通过 XML 文件定义新网络,第 3 行命令用于设置自动启动,第 4 行命令用于启动这个开放式网络。

第 5 行命令的输出显示新网络创建成功了。

接下来执行的命令如下:

```
6 # virsh net - info open1
  Name:          open1
  UUID:          55dfc4da - dccf - 47b7 - a2b9 - d2f5c130804c
  Active:        yes
  Persistent:    yes
  Autostart:     yes
  Bridge:        virbr1
```

```
7 # virsh net-dumpxml open1
  <network>
    <name>open1</name>
    <uuid>55dfc4da-dccf-47b7-a2b9-d2f5c130804c</uuid>
    <forward mode='open'/>
    <bridge name='virbr1' stp='on' delay='0'/>
    <mac address='52:54:00:16:a8:57'/>
    <domain name='open1'/>
    <ip address='192.168.100.1' netmask='255.255.255.0'>
      <dhcp>
        <range start='192.168.100.128' end='192.168.100.254'/>
      </dhcp>
    </ip>
  </network>
```

通过第 6 行、第 7 行命令可以查看新网络的信息。

接下来执行的命令如下：

```
8 # ip address
1: lo: <LOOPBACK,UP,LOWER_UP> mtu 65536 qdisc noqueue state UNKNOWN group default qlen 1000
    link/loopback 00:00:00:00:00:00 brd 00:00:00:00:00:00
    inet 127.0.0.1/8 scope host lo
       valid_lft forever preferred_lft forever
    inet6 ::1/128 scope host
       valid_lft forever preferred_lft forever
2: ens32: <BROADCAST,MULTICAST,UP,LOWER_UP> mtu 1500 qdisc fq_codel state UP group default qlen 1000
    link/ether 00:0c:29:f7:6b:c8 brd ff:ff:ff:ff:ff:ff
    inet 192.168.1.48/24 brd 192.168.1.255 scope global noprefixroute ens32
       valid_lft forever preferred_lft forever
    inet6 fe80::20c:29ff:fef7:6bc8/64 scope link
       valid_lft forever preferred_lft forever
3: virbr0: <NO-CARRIER,BROADCAST,MULTICAST,UP> mtu 1500 qdisc noqueue state DOWN group default qlen 1000
    link/ether 52:54:00:e0:41:ac brd ff:ff:ff:ff:ff:ff
    inet 192.168.122.1/24 brd 192.168.122.255 scope global virbr0
       valid_lft forever preferred_lft forever
4: virbr0-nic: <BROADCAST,MULTICAST> mtu 1500 qdisc fq_codel master virbr0 state DOWN group default qlen 1000
    link/ether 52:54:00:e0:41:ac brd ff:ff:ff:ff:ff:ff
5: virbr1: <NO-CARRIER,BROADCAST,MULTICAST,UP> mtu 1500 qdisc noqueue state DOWN group default qlen 1000
    link/ether 52:54:00:16:a8:57 brd ff:ff:ff:ff:ff:ff
    inet 192.168.100.1/24 brd 192.168.100.255 scope global virbr1
       valid_lft forever preferred_lft forever
```

```
6: virbr1-nic: <BROADCAST,MULTICAST> mtu 1500 qdisc fq_codel master virbr1 state DOWN
group default qlen 1000
    link/ether 52:54:00:16:a8:57 brd ff:ff:ff:ff:ff:ff
```

从第 8 行命令的输出可以看出：新增加了 1 个名为 virbr1 的网桥，其 IP 地址是 192.168.100.1/24，它有一个名为 virbr1-nic 的子接口。

接下来执行的命令如下：

```
9 # iptables -L -n -v
  Chain INPUT (policy ACCEPT 963 packets, 70270 Bytes)
   target     prot opt in     out    source       destination
   ACCEPT     udp  --  virbr0  *     0.0.0.0/0    0.0.0.0/0    udp dpt:53
   ACCEPT     tcp  --  virbr0  *     0.0.0.0/0    0.0.0.0/0    tcp dpt:53
   ACCEPT     udp  --  virbr0  *     0.0.0.0/0    0.0.0.0/0    udp dpt:67
   ACCEPT     tcp  --  virbr0  *     0.0.0.0/0    0.0.0.0/0    tcp dpt:67

  Chain FORWARD (policy ACCEPT 0 packets, 0 Bytes)
   target prot opt in     out     source          destination
   ACCEPT all  --  *      virbr0  0.0.0.0/0       192.168.122.0/24    ctstate
RELATED,ESTABLISHED
   ACCEPT all  --  virbr0 *       192.168.122.0/24   0.0.0.0/0
   ACCEPT all  --  virbr0 virbr0  0.0.0.0/0       0.0.0.0/0
   REJECT all  --  *      virbr0  0.0.0.0/0       0.0.0.0/0           reject-with
icmp-port-unreachable
   REJECT all  --  virbr0 *       0.0.0.0/0       0.0.0.0/0           reject-with
icmp-port-unreachable

  Chain OUTPUT (policy ACCEPT 508 packets, 62659 Bytes)
   target prot opt in     out     source      destination
   ACCEPT udp  --  *      virbr0  0.0.0.0/0   0.0.0.0/0    udp dpt:68
```

与路由模式不同，libvirt 不会为开放模式的网络设置防火墙规则，所以不会在第 9 行命令的输出中看到与 192.168.100.0/24 和 virbr1 相关的策略。

接下来执行的命令如下：

```
10 # virsh edit centos6.10
   ...
   <interface type='network'>
     <mac address='52:54:00:27:5f:c9'/>
     <source network='open1'/>
     <model type='virtio'/>
     <address type='pci' domain='0x0000' bus='0x00' slot='0x03' function='0x0'/>
   </interface>
   ...
```

```
11 # virsh start centos6.10
```

第 10 行命令用于修改虚拟机 centos6.10 的网卡属性，使其连接到 open1 这个开放的网络。

第 11 行命令用于启动虚拟机。

默认情况下，即使为 open1 网络配置 DHCP 地址范围（192.168.100.128 至 192.168.100.254），虚拟机 centos6.10 也无法得到 IP 地址，如图 5-34 所示。

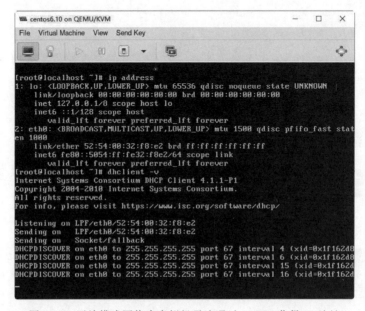

图 5-34　开放模式网络中虚拟机无法通过 DHCP 获得 IP 地址

接下来执行的命令如下：

```
12 # virsh net-dhcp-leases open1
   Expiry Time    MAC address    Protocol    IP address    Hostname    Client ID or DUID
   ----------------------------------------------------------------------------------
```

第 12 行命令的输出也显示了这个名为 open1 的网络并没有为虚拟机分配 IP 地址。

出现这种现象的原因是：libvirt 并没有为 192.168.100.0/24 和 virbr1 设置允许 DHCP 数据的规则，所以虚拟机发出的 DHCPDISCOVER 包并没有被宿主机上的 DNSMASQ 组件收到。

如果希望为 open1 中的虚拟机自动分配 IP 地址，就需要手工添加防火墙规则。命令如下：

```
13 #firewall-cmd --list-all
   public (active)
     target: default
     icmp-block-inversion: no
     interfaces: ens32
     sources:
     services: cockpit dhcpv6-client ssh
     ports:
     protocols:
     masquerade: no
     forward-ports:
     source-ports:
     icmp-blocks:
     rich rules:

14 #firewall-cmd --add-interface=virbr1

15 #firewall-cmd --add-service=dhcp

16 #firewall-cmd --list-all
   public (active)
     target: default
     icmp-block-inversion: no
     interfaces: ens32 virbr1
     sources:
     services: cockpit dhcp dhcpv6-client ssh
     ports:
     protocols:
     masquerade: no
     forward-ports:
     source-ports:
     icmp-blocks:
     rich rules:
```

RHEL/CentOS 8 推荐使用 firewall 的 firewall-cmd 命令来管理防火墙。通过第 13 行命令来获得当前防火墙配置的概要信息。

第 14 行命令向 public 区域添加了一个接口 virbr1。第 15 行命令又允许 DHCP 服务，这样将允许来自 open1 网络的与 DHCP 相关的数据包到达宿主机，DNSMASQ 收到后就会为其分配 IP 地址。

接下来执行的命令如下：

```
17 #virsh net-dhcp-leases open1
   Expiry Time        MAC address        Protocol IP address      Hostname
Client ID or DUID
------------------------------------------------------------------------
   2020-12-01 20:56:03    52:54:00:32:f8:e2    ipv4    192.168.100.182/24
```

第 17 行命令的输出显示：向虚拟机分配的 IP 地址是 192.168.100.182/24。

后续的配置与路由模式类似，这里就不再赘述了。

> 提示：如果需要使防火墙配置永久生效，还需要使用--permanent 选项来执行 firewall-cmd 命令。

在配置路由模式时，libvirt 会为其配置防火墙规则，例如允许 DHCP、DNS 等流量，而在配置开放模式时，libvirt 不会为此虚拟网络生成任何 iptables 规则，这样就需要用户自行配置了。

5.10 实现多 VLAN 支持

虚拟局域网(Virtual LAN,VLAN)是物理网络中的逻辑网络，它提高了网络管理的安全性与灵活性。通过使用 VLAN 标记，可以将每个以太网端口都视为包含许多逻辑端口。通过合理的规划与配置，可以在一台 KVM 宿主机中运行属于不同 VLAN 的虚拟机，如图 5-35 所示。

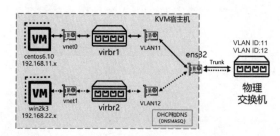

图 5-35　同一个宿主机中可以运行属于多个不同 VLAN 的虚拟机

图 5-35 中的物理网络有两个 VLAN，其 ID 分别是 11 和 12。虚拟机 centos6.10 属于 ID 为 11 的 VLAN，虚拟机 Win2k3 则属于 ID 为 12 的 VLAN。

首先，需要在宿主机现有接口(例如以太网卡、绑定网卡或网桥设备)上创建 VLAN 接口，在图 5-35 中这个接口是物理网卡 ens32。在 ens32 上创建两个使用不同 VLAN ID 的虚拟网络接口 VLAN11 和 VLAN12，ens32 被称为 VLAN11、VLAN12 的"父"接口。

然后创建两个网桥 virbr1 和 virbr2。其中 virbr1 的子接口是 VLAN11，virbr2 的子接口是 VLAN12。

将虚拟机 centos6.10 桥接到 virbr1，而虚拟机 Win2k3 则需要桥接到 virbr2。

在宿主机所在的物理网络中也需要进行设置，主要就是将宿主机网卡 ens32 所连接的交换机端口配置为 trunk 模式。

5.10.1 创建支持 VLAN 的网络接口

Linux 通过在内核中加载 8021q 模块实现 VLAN 功能。在 RHEL/CentOS 8 中，默认情况下会根据需要自动加载 8021q 模块。在其他的 Linux 发版本中，可能需要使用 modprobe 命令来手工加载。

在 RHEL/CentOS 8 中，既可以通过 iproute2、NetworkManager、Cockpit 等多种工具来管理 VLAN，也可以直接编辑 /etc/sysconfig/network-scripts/ 目录下的网络接口脚本文件实现 VLAN 的管理。由于 iproute2 中的 ip 命令所创建的 VLAN 不是永久性的，所以不推荐使用。下面，我们通过 nmcli 和 Cockpit 这两种工具来创建和管理 VLAN，命令如下：

```
1 # ip address
  1: lo: <LOOPBACK,UP,LOWER_UP> mtu 65536 qdisc noqueue state UNKNOWN group default qlen 1000
     link/loopback 00:00:00:00:00:00 brd 00:00:00:00:00:00
     inet 127.0.0.1/8 scope host lo
        valid_lft forever preferred_lft forever
     inet6 ::1/128 scope host
        valid_lft forever preferred_lft forever
  2: ens32: <BROADCAST,MULTICAST,UP,LOWER_UP> mtu 1500 qdisc fq_codel state UP group default qlen 1000
     link/ether 00:0c:29:b3:73:36 brd ff:ff:ff:ff:ff:ff
     inet 192.168.1.231/24 brd 192.168.1.255 scope global noprefixroute ens32
        valid_lft forever preferred_lft forever
     inet6 fe80::20c:29ff:feb3:7336/64 scope link
        valid_lft forever preferred_lft forever
  3: virbr0: <NO-CARRIER,BROADCAST,MULTICAST,UP> mtu 1500 qdisc noqueue state DOWN group default qlen 1000
     link/ether 52:54:00:fd:b2:60 brd ff:ff:ff:ff:ff:ff
     inet 192.168.122.1/24 brd 192.168.122.255 scope global virbr0
        valid_lft forever preferred_lft forever
  4: virbr0-nic: <BROADCAST,MULTICAST> mtu 1500 qdisc fq_codel master virbr0 state DOWN group default qlen 1000
     link/ether 52:54:00:fd:b2:60 brd ff:ff:ff:ff:ff:ff

2 # ls /etc/sysconfig/network-scripts/
  ifcfg-ens32

3 # cat /etc/sysconfig/network-scripts/ifcfg-ens32
  TYPE = Ethernet
  NAME = ens32
  DEVICE = ens32
  ONBOOT = yes
  IPADDR = 192.168.1.231
  PREFIX = 24
  GATEWAY = 192.168.1.1
  DNS1 = 8.8.8.8
```

通过第 1 行命令查看当前网络接口的地址信息。

第 2 行命令显示在 /etc/sysconfig/network-scripts/ 目录中仅包含 ens32 的接口配置文件 ifcfg-ens32。通过第 3 行命令查看文件的内容，这是 1 个精简后的配置文件，仅保留必要的设置值。

接下来执行的命令如下：

```
4 # nmcli connection add type vlan con-name vlan11 ifname vlan11 vlan.parent ens32 vlan.id 11
  Connection 'vlan11' (76131bf6-2889-4e44-9a4a-90281bd2e22a) successfully added.

5 # nmcli connection modify vlan11 ipv4.addresses '192.168.11.231/24' ipv4.gateway 192.168.11.1 ipv4.method manual

6 # nmcli connection up vlan11
  Connection successfully activated (D-Bus active path: /org/freedesktop/NetworkManager/ActiveConnection/10)
```

我们先通过 NetworkManager 的命令行工具 nmcli 来创建新连接。

第 4 行命令创建了一个新的连接，通过 type 指定类型为 vlan，通过 con-name 指定新连接的名称为 vlan11，通过 ifname 指定的接口名称也是 vlan11，vlan.parent 指定父接口是 ens32，vlan.id 指定 VLAN 的 ID 是 11。

第 5 行命令用于设置 VLAN 网络接口的 IPv4 地址、网关和分配地址的模式。

第 6 行命令激活这个连接。

接下来执行的命令如下：

```
7 # ip -d address show vlan11
  10: vlan11@ens32: <BROADCAST,MULTICAST,UP,LOWER_UP> mtu 1500 qdisc noqueue state UP group default qlen 1000
      link/ether 00:0c:29:b3:73:36 brd ff:ff:ff:ff:ff:ff promiscuity 0 minmtu 0 maxmtu 65535
      vlan protocol 802.1Q id 11 <REORDER_HDR> numtxqueues 1 numrxqueues 1 gso_max_size 65536 gso_max_segs 65535
      inet 192.168.11.231/24 brd 192.168.11.255 scope global noprefixroute vlan11
         valid_lft forever preferred_lft forever
      inet6 fe80::1db2:9a6c:6979:894e/64 scope link noprefixroute
         valid_lft forever preferred_lft forever

8 # ip -d address show ens32
  2: ens32: <BROADCAST,MULTICAST,UP,LOWER_UP> mtu 1500 qdisc fq_codel state UP group default qlen 1000
      link/ether 00:0c:29:b3:73:36 brd ff:ff:ff:ff:ff:ff promiscuity 0 minmtu 46 maxmtu 16110 numtxqueues 1 numrxqueues 1 gso_max_size 65536 gso_max_segs 65535
      inet 192.168.1.231/24 brd 192.168.1.255 scope global noprefixroute ens32
         valid_lft forever preferred_lft forever
      inet6 fe80::20c:29ff:feb3:7336/64 scope link
         valid_lft forever preferred_lft forever
```

在第 7 行命令中使用了 -d 选项，会显示连接的详细信息。自动生成的连接名称"vlan11@ens32"很直观，表示 vlan11 是在 ens32 之上实现的。除了 IP 地址等基本信息之外，"vlan protocol 802.1Q id 11"显示了 VLAN 的协议和 ID 号。

对比一下第 8 行命令的输出，会发现新创建的 vlan11"借用"了 ens32 的 MAC 地址"00:0c:29:b3:73:36"。

接下来执行的命令如下：

```
9 # cat /etc/sysconfig/network-scripts/ifcfg-vlan11
  VLAN = yes
  TYPE = Vlan
  PHYSDEV = ens32
  VLAN_ID = 11
  REORDER_HDR = yes
  GVRP = no
  MVRP = no
  HWADDR =
  PROXY_METHOD = none
  BROWSER_ONLY = no
  BOOTPROTO = none
  DEFROUTE = yes
  IPV4_FAILURE_FATAL = no
  IPV6INIT = yes
  IPV6_AUTOCONF = yes
  IPV6_DEFROUTE = yes
  IPV6_FAILURE_FATAL = no
  IPV6_ADDR_GEN_MODE = stable-privacy
  NAME = vlan11
  UUID = 76131bf6-2889-4e44-9a4a-90281bd2e22a
  DEVICE = vlan11
  ONBOOT = yes
  IPADDR = 192.168.11.231
  PREFIX = 24
  GATEWAY = 192.168.11.1
```

NetworkManager 创建新的 VLAN 接口，还会在/etc/sysconfig/network-scripts/目录中创建网络接口配置文件，第 9 行命令显示了这个配置文件的内容，其中最重要的属性有以下几种。

(1) VLAN＝yes。

(2) TYPE＝Vlan。

(3) PHYSDEV＝ens32。

(4) VLAN_ID＝11。

(5) DEVICE＝vlan11。

(6) ONBOOT＝yes。

提示：可以通过 man nm-settings-ifcfg-rh 来获得 /etc/sysconfig/network-scripts/ifcfg-* 设置的详细说明。

接下来执行的命令如下：

```
10 #lsmod | grep 8021q
   8021q    40960   0
   garp     16384   1 8021q
   mrp      20480   1 8021q
```

当激活 VLAN 网络接口时，NetworkManager 会自动将 8021q 模块加载到内核中，可以从第 10 行命令的输出中看到这个模块。

相对于 nmcli，通过 Cockpit 来创建新的 VLAN 接口特别简单。

（1）在 Cockpit 左边导航中单击 Networking，会显示当前的所有网络接口，如图 5-36 所示。

图 5-36　Cockpit 中的网络接口信息

（2）单击 Add VLAN 按钮。

（3）在 VLAN Setting 对话框中，选择要为其创建 VLAN 的物理接口 ens32，将 VLAN ID 设置为 12，将接口名称设置为 vlan12，然后单击 Apply 按钮完成创建，如图 5-37 所示。

（4）新 VLAN 创建完毕，如图 5-38 所示。单击 Configure IP 链接配置网络设置。

（5）在 IPv4 Setting 对话框中设置 IP 地址及缺省网关等信息，然后单击 Apply 按钮保存配置，如图 5-39 所示。

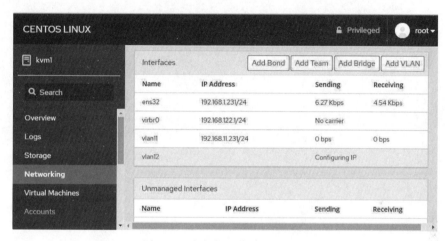

图 5-37 设置新 VLAN 网络接口的信息

图 5-38 包括新网络接口的列表

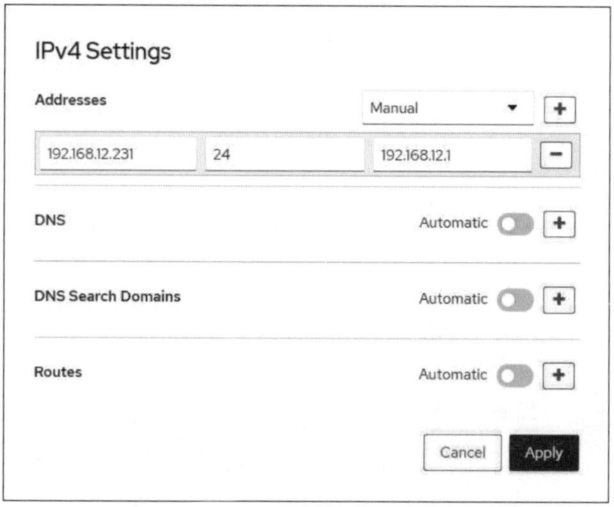

图 5-39 为新的网络接口设置 IP 地址等信息

Cockpit 还会在 /etc/sysconfig/network-scripts/ 目录中创建网络接口配置文件 ifcfg-vlan12，命令如下：

```
11 # cat /etc/sysconfig/network-scripts/ifcfg-vlan12
   VLAN = yes
   TYPE = Vlan
   PHYSDEV = ens32
   VLAN_ID = 12
   REORDER_HDR = yes
   GVRP = no
   MVRP = no
   HWADDR =
   PROXY_METHOD = none
   BROWSER_ONLY = no
   BOOTPROTO = none
   DEFROUTE = yes
   IPV4_FAILURE_FATAL = no
   IPV6INIT = yes
   IPV6_AUTOCONF = yes
   IPV6_DEFROUTE = yes
   IPV6_FAILURE_FATAL = no
   IPV6_ADDR_GEN_MODE = stable-privacy
   NAME = vlan12
   UUID = ae59cf5b-f088-4ba6-9f20-7eb2def383fe
   DEVICE = vlan12
   ONBOOT = yes
   IPADDR = 192.168.12.231
   PREFIX = 24
   GATEWAY = 192.168.12.1
   PEERDNS = no
   PEERROUTES = no
```

提示：也可以通过在 /etc/sysconfig/network-scripts/ 目录中创建新的配置文件的方法来创建新的支持的 VLAN 接口。

5.10.2 创建使用 VLAN 网络接口的网桥

有了两个支持 VLAN 的网络接口 VLAN11 和 VLAN12，根据图 5-35 所示的拓扑，我们需要在宿主机上创建两个网桥 virbr1 和 virbr2。VLAN11 是 virbr1 的子接口，VLAN12 是 virbr2 的子接口。

与前面创建网桥的方法类似，最简单的方法是通过 Cockpit 来创建。

（1）在 Cockpit 左边导航中单击 Networking，会显示当前的所有网络接口，如图 5-40 所示。

(2)单击 Add Bridge 按钮。

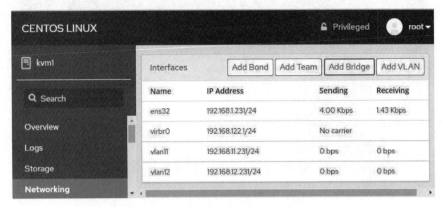

图 5-40　Cockpit 中的网络接口信息

(3)在 Bridge Setting 对话框中,将网桥名设置为 virbr1,选择子接口 vlan11,然后单击 Apply 按钮完成创建,如图 5-41 所示。

图 5-41　设置新网桥的信息

(4)类似地,再创建使用 VLAN12 的新网桥 virbr2,最后的结果如图 5-42 所示。

5.10.3　配置虚拟机使用 VLAN

将虚拟机 centos6.10 桥接到 virbr1,而虚拟机 win2k3 则需要桥接到 virbr2。如果使用的是 Cockpit,则可在 Interface Type 选择 Bridge to LAN,在 Source 中选择适当的网桥,如图 5-43 所示。

图 5-42　包括新网桥的列表

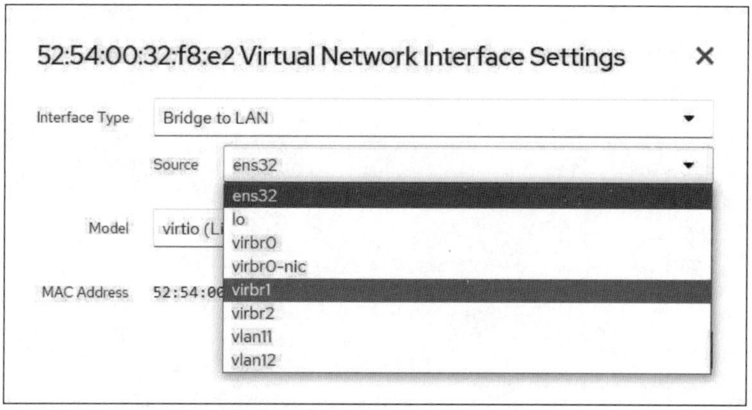

图 5-43　使用 Cockpit 更新虚拟机的网络配置

也可以通过 virsh 的 edit 等子命令来修改虚拟机的配置文件，最重要的是＜source bridge＝'virbr1'/＞这一行的设置，配置文件的内容如下：

```
< interface type = 'bridge'>
  < mac address = '52:54:00:32:f8:e2'/>
  < source bridge = 'virbr1'/>
  < model type = 'virtio'/>
  < address type = 'pci' domain = '0x0000' bus = '0x00' slot = '0x03'
function = '0x0'/>
</ interface >
```

在此实验环境中，我们可以使用 Wireshark 等协议分析工具在宿主机外面的网络中抓到来自虚拟机而且打有 VLAN 标记的数据包，如图 5-44 所示。

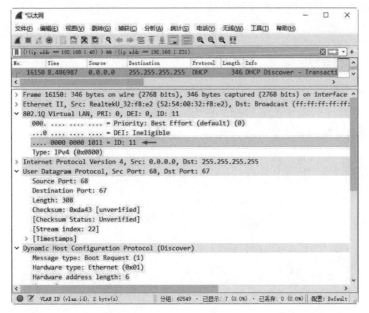

图 5-44　来自虚拟机 centos6.10 并带有 VLAN 标记的数据包

5.11　通过网络过滤器提高安全性

网络过滤器（Network Filter）是 libvirt 的一个子系统，可以通过在其中配置过滤规则来对每个虚拟机网络接口的流量进行控制，从而提高安全性。libvirt 会将这些过滤规则转换为宿主机上的防火墙规则，在虚拟机启动时自动添加与其网络接口有关的防火墙规则，而当其关闭时也会自动删除这些规则。

与虚拟机内部的防火墙不同，网络过滤规则应用于宿主机之上，所以无法从虚拟机内部规避这些过滤规则，因此从虚拟机用户的角度来看，这些规则是透明的、强制性的、无法"旁路"的，如图 5-45 所示。

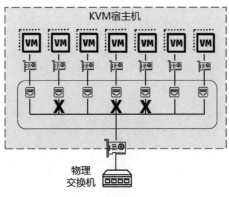

图 5-45　网络过滤器原理

既可以针对特定虚拟机的某个网络接口配置网络过滤规则,也可以让多个虚拟机使用相同的过滤规则。

5.11.1 网络过滤器基本原理

下面通过一个简单的实验来了解网络过滤器的基本原理。

在实验中,创建一个阻止所有 ICMP 流量的过滤器,然后将其应用于虚拟机 centos6.10。

首先,查看虚拟机 centos6.10 的网络连通性,命令如下:

```
1 # virsh domifaddr centos6.10
    Name       MAC address          Protocol     Address
   -------------------------------------------------------------
    vnet0      52:54:00:27:5f:c9    ipv4         192.168.122.142/24

2 # ping -c 2 192.168.122.142
   PING 192.168.122.142 (192.168.122.142) 56(84) Bytes of data.
   64 Bytes from 192.168.122.142: icmp_seq=1 ttl=64 time=0.939 ms
   64 Bytes from 192.168.122.142: icmp_seq=2 ttl=64 time=1.20 ms

   --- 192.168.122.142 ping statistics ---
   2 packets transmitted, 2 received, 0% packet loss, time 3ms
   rtt min/avg/max/mdev = 0.939/1.071/1.204/0.136 ms

3 # iptables -L -n
   Chain INPUT (policy ACCEPT)
   target     prot opt source           destination
   ACCEPT     udp  --  0.0.0.0/0        0.0.0.0/0        udp dpt:53
   ACCEPT     tcp  --  0.0.0.0/0        0.0.0.0/0        tcp dpt:53
   ACCEPT     udp  --  0.0.0.0/0        0.0.0.0/0        udp dpt:67
   ACCEPT     tcp  --  0.0.0.0/0        0.0.0.0/0        tcp dpt:67

   Chain FORWARD (policy ACCEPT)
   target     prot opt source           destination
   ACCEPT     all  --  0.0.0.0/0        192.168.122.0/24   ctstate RELATED,ESTABLISHED
   ACCEPT     all  --  192.168.122.0/24 0.0.0.0/0
   ACCEPT     all  --  0.0.0.0/0        0.0.0.0/0
   REJECT     all  --  0.0.0.0/0        0.0.0.0/0        reject-with icmp-port-unreachable
   REJECT     all  --  0.0.0.0/0        0.0.0.0/0        reject-with icmp-port-unreachable

   Chain OUTPUT (policy ACCEPT)
   target     prot opt source           destination
   ACCEPT     udp  --  0.0.0.0/0        0.0.0.0/0        udp dpt:68
```

通过第 1 行命令的输出可以看出:虚拟机 centos6.10 网络接口在宿主机上的名称为 vnet0,IP 地址为 192.168.122.142。

第 2 行命令的输出显示了可以在宿主机上 ping 通虚拟机，这说明虚拟机操作系统的防火墙没有阻止 ICMP 的 echo 请求与响应。

第 3 行命令的输出显示了宿主机上当前防火墙的配置，其中并没有针对 vnet0 或 192.168.122.142 的特别设置。

接下来执行的命令如下：

```
4 # vi new.xml
  <filter name='blockicmp'>
    <rule action='drop' direction='in'>
      <icmp/>
    </rule>
  </filter>

5 # virsh nwfilter-define new.xml
  Network filter blockicmp defined from new.xml

6 # virsh nwfilter-list
  UUID                                   Name
  ------------------------------------------------------------------
  4b69faf1-fac2-4efe-9f30-d8bd12d0de87   allow-arp
  bdaaee09-70ff-4bff-a0c5-f1d9313c19df   allow-dhcp
  93a8a4ca-2714-4d93-8df3-1610bf0eac06   allow-dhcp-server
  fe8d7784-7b2d-4d8e-9aa4-642ca8e282f4   allow-incoming-ipv4
  34ab7687-59b0-414e-9294-4884dfe7c111   allow-ipv4
  737777d6-1115-425a-9063-defc0e14b614   blockicmp
  da713eaa-e490-4d61-8489-b62110795ce4   clean-traffic
  489f315a-4ed5-4fa3-8336-9acebc1dee83   clean-traffic-gateway
  843b6276-68a2-4bd8-8a31-76efe4ee7526   no-arp-ip-spoofing
  0c6a9780-f1e5-4181-a758-c1aa58321ff0   no-arp-mac-spoofing
  e9dd8d7d-0fde-4cb5-9917-b192d761782b   no-arp-spoofing
  76da0b46-fc7c-4991-b421-f18210a4887a   no-ip-multicast
  724ec182-c6fe-4289-a1d3-8b0b4821a897   no-ip-spoofing
  d8a940e2-ebdf-4ffd-9c85-75023362829f   no-mac-broadcast
  5036c658-6bea-4299-aff4-0bd53a9048dd   no-mac-spoofing
  a4b3fc68-629e-45a3-b42a-73c6b1b19879   no-other-l2-traffic
  397ccccf-c425-4c93-98f9-35fb3870c61c   no-other-rarp-traffic
  746cf6aa-ff87-497d-a0f6-89352bde27b0   qemu-announce-self
  a3b67cb4-e977-46db-a084-24e9b9e74b63   qemu-announce-self-rarp
```

为了创建网络过滤器，需要创建一个 XML 格式的文件，其内容如第 4 行命令所示。<filter name='blockicmp'>指定新的过滤器的名称为 blockicmp，并且通过<rule action='drop' direction='in'>来创建一个规则，action='drop'表示这是一个阻止规则，direction='in'表示只针对入站的流量，<icmp/>用于指定协议为 icmp。

第 5 行命令使用 virsh 的 nwfilter-define 子命令来创建网络过滤器。

第 6 行命令的输出显示了当前系统中网络过滤器的列表。除了新的网络过滤器 blockicmp 之外，还会看到多个预安装的过滤器。

提示：virsh 有多个与网络过滤器有关的并以 nwfilter-开头的子命令，nwfilter 是 Network Filter 的缩写。

接下来执行的命令如下：

```
7 # virsh shutdown centos6.10

8 # virsh edit centos6.10
    ...
    < interface type = 'network'>
      < mac address = '52:54:00:27:5f:c9'/>
      < source network = 'default'/>
      < model type = 'virtio'/>
      < address type = 'pci' domain = '0x0000' bus = '0x00' slot = '0x03' function = '0x0'/>
      < filterref filter = 'blockicmp'/>
    </interface>
    ...

9 # virsh start centos6.10
```

第 7 行命令用于关闭虚拟机。

第 8 行命令用于修改虚拟机的配置文件。通过< filterref filter = 'blockicmp'/>将虚拟机的网络接口与网络过滤器 blockicmp 相关联。

第 9 行命令用于启动虚拟机。

接下来执行的命令如下：

```
10 # ping - c 2 192.168.122.142
    PING 192.168.122.142 (192.168.122.142) 56(84) Bytes of data.

    --- 192.168.122.142 ping statistics ---
    2 packets transmitted, 0 received, 100% packet loss, time 27ms

11 # ssh 192.168.122.142
    root@192.168.122.142's password:
    Last login: Thu Dec 17 11:29:33 2020 from 192.168.122.1
    [root@localhost ~] # exit
    logout
    Connection to 192.168.122.142 closed.
```

虚拟机启动之后，可在宿主机上测试其与虚拟机的连通性。从第 10 行命令的输出可以看出：无法在宿主机上 ping 通虚拟机，这是由于网络过滤器 blockicmp 阻止了所有的 ICMP 流量。

不过，由于网络过滤器 blockicmp 仅仅阻止了 ICMP 流量，所以可以通过 SSH 访问这台虚拟机，如第 11 行命令的输出所示。

接下来执行的命令如下：

```
12 # iptables -L -n
   Chain INPUT (policy ACCEPT)
   target              prot opt source          destination
   libvirt-host-in     all  --  0.0.0.0/0       0.0.0.0/0
   ACCEPT              udp  --  0.0.0.0/0       0.0.0.0/0       udp dpt:53
   ACCEPT              tcp  --  0.0.0.0/0       0.0.0.0/0       tcp dpt:53
   ACCEPT              udp  --  0.0.0.0/0       0.0.0.0/0       udp dpt:67
   ACCEPT              tcp  --  0.0.0.0/0       0.0.0.0/0       tcp dpt:67

   Chain FORWARD (policy ACCEPT)
   target              prot opt source          destination
   libvirt-in          all  --  0.0.0.0/0       0.0.0.0/0
   libvirt-out         all  --  0.0.0.0/0       0.0.0.0/0
   libvirt-in-post     all  --  0.0.0.0/0       0.0.0.0/0
   ACCEPT              all  --  0.0.0.0/0       192.168.122.0/24    ctstate
RELATED,ESTABLISHED
   ACCEPT              all  --  192.168.122.0/24  0.0.0.0/0
   ACCEPT              all  --  0.0.0.0/0       0.0.0.0/0
   REJECT              all  --  0.0.0.0/0       0.0.0.0/0       reject-with
icmp-port-unreachable
   REJECT              all  --  0.0.0.0/0       0.0.0.0/0       reject-with
icmp-port-unreachable

   Chain OUTPUT (policy ACCEPT)
   target     prot opt source           destination
   ACCEPT     udp  --  0.0.0.0/0        0.0.0.0/0       udp dpt:68

   Chain libvirt-in (1 references)
   target     prot opt source      destination
   FI-vnet0   all  --  0.0.0.0/0  0.0.0.0/0   [goto] PHYSDEV match -- physdev-in vnet0

   Chain libvirt-out (1 references)
   target     prot opt source      destination
   FO-vnet0   all  --  0.0.0.0/0  0.0.0.0/0   [goto] PHYSDEV match -- physdev-out vnet0 --
physdev-is-bridged

   Chain libvirt-in-post (1 references)
   target     prot opt source      destination
   ACCEPT     all  --  0.0.0.0/0  0.0.0.0/0   PHYSDEV match -- physdev-in vnet0
```

```
Chain libvirt-host-in (1 references)
target     prot opt source          destination
HI-vnet0   all  --  0.0.0.0/0       0.0.0.0/0       [goto] PHYSDEV match -- physdev-in vnet0

Chain FO-vnet0 (1 references)
target     prot opt source          destination
DROP       icmp --  0.0.0.0/0       0.0.0.0/0

Chain FI-vnet0 (1 references)
target     prot opt source          destination
DROP       icmp --  0.0.0.0/0       0.0.0.0/0

Chain HI-vnet0 (1 references)
target     prot opt source          destination
DROP       icmp --  0.0.0.0/0       0.0.0.0/0
```

第12行命令的输出显示了当前宿主机上的防火墙的配置,增加了多个自定义的链。其中 libvirt-host-in、libvirt-in、libvirt-in-post、libvirt-out 是当 libvirt 第一次加载网络过滤器时自动创建的,而 FI-vnet0、FO-vnet0、HI-vnet0 是在初始化虚拟机 centos6.10 的网络接口时创建的(当前虚拟机 centos6.10 在宿主机网卡上的名称为 vnet0)。

libvirt 通过调用系统的 iptables、ip6tables 和 ebtables 命令完成防火墙的配置,这可以通过查看系统日志/var/log/messages 得到印证。

可以将一个网络过滤器与多台虚拟机相关联,从而实现共享配置。例如:修改虚拟机 win2k3 的配置文件,并且使用网络过滤器 blockicmp,这样也可以阻止它的 ICMP 流量。

下面清除实验环境,命令如下:

```
13 # virsh shutdown centos6.10

14 # virsh edit centos6.10
    删除<filterref filter='blockicmp'/>

15 # virsh nwfilter-undefine blockicmp
   Network filter blockicmp undefined
```

只能删除未使用的网络过滤器。第13行命令先关闭虚拟机,然后编辑配置文件,删除<filterref filter='blockicmp'/>。最后使用第15行命令取消网络过滤器的定义。

5.11.2 网络过滤器的管理工具

目前还不能通过 Cockpit、virt-manager 来管理网络过滤器,所以我们主要通过 virsh 中 nwfilter-开头的子命令进行维护。

(1) nwfilter-define:根据 XML 文件定义或更新网络过滤器。

(2) nwfilter-dumpxml：输出 XML 格式的网络过滤器信息。

(3) nwfilter-edit：编辑网络过滤器的 XML 配置文件。

(4) nwfilter-list：列出网络过滤器清单。

(5) nwfilter-undefine：取消网络过滤器的定义。

(6) nwfilter-binding-create：根据 XML 文件创建网络过滤器的绑定。

(7) nwfilter-binding-delete：删除网络过滤器的绑定。

(8) nwfilter-binding-dumpxml：输出 XML 格式的网络过滤器所绑定的信息。

(9) nwfilter-binding-list：列出网络过滤器绑定的清单。

执行的命令如下：

```
1 #virsh help filter
 Network Filter (help keyword 'filter'):
   nwfilter-define            define or update a network filter from an XML file
   nwfilter-dumpxml           network filter information in XML
   nwfilter-edit              edit XML configuration for a network filter
   nwfilter-list              list network filters
   nwfilter-undefine          undefine a network filter
   nwfilter-binding-create    create a network filter binding from an XML file
   nwfilter-binding-delete    delete a network filter binding
   nwfilter-binding-dumpxml   network filter information in XML
   nwfilter-binding-list      list network filter bindings

2 #virsh nwfilter-define --help
   NAME
     nwfilter-define - define or update a network filter from an XML file

   SYNOPSIS
     nwfilter-define <file>

   DESCRIPTION
     Define a new network filter or update an existing one.

   OPTIONS
     [--file] <string>  file containing an XML network filter description
```

5.11.3　预安装的网络过滤器

在 RHEL/CentOS 8 中安装虚拟化组件时，会自动安装一些网络过滤器。与自定义的网络过滤器一样，都保存在 /etc/libvirt/nwfilter/ 目录中。

这些过滤器的命名很直观，可以从名称猜测出其大概的用途。详细的描述如表 5-14 所示。

表 5-14 预安装的网络过滤器

名称	描述
no-arp-spoofing	防止虚拟机欺骗 ARP 流量；该过滤器仅允许 ARP 请求和答复消息，并强制这些数据包包含虚拟机的 MAC 和 IP 地址
allow-arp	允许双向 ARP 流量
allow-ipv4	允许双向 IPv4 流量
allow-ipv6	允许双向 IPv6 流量
allow-incoming-ipv4	允许传入的 IPv4 流量
allow-incoming-ipv6	允许传入的 IPv6 流量
allow-dhcp	允许虚拟机通过任何 DHCP 服务器获得 IP 地址
allow-dhcpv6	与 allow-dhcp 类似，但用于 DHCPv6
allow-dhcp-server	允许虚拟机从指定的 DHCP 服务器获得 IP 地址。必须对此过滤器提供 DHCP 服务器的 IPv4 地址。变量的名称必须为 DHCPSERVER
allow-dhcpv6-server	与 allow-dhcp-server 类似，但用于 DHCPv6
no-ip-spoofing	防止虚拟机发送源 IP 地址与数据包中的源 IP 地址不同的 IPv4 数据包
no-ipv6-spoofing	类似于 no-ip-spoofing，但用于 IPv6
no-ip-multicast	防止虚拟机发送 IP 组播报文
no-ipv6-multicast	类似于 no-ip-multicast，但用于 IPv6
clean-traffic	防止 MAC、IP 和 ARP 欺骗。该过滤器引用其他几个过滤器作为构建块

上述大多数网络过滤器只是构建块，需要与其他网络过滤器结合使用以提供有用的网络流量过滤。下面我们分析一下网络过滤器 clean-traffic，它引用了多个其他的网络过滤器，代码如下：

```
1 # virsh nwfilter-dumpxml clean-traffic
  <filter name = 'clean-traffic' chain = 'root'>
    <uuid>1a4aadd7-62c0-47fe-a91b-6becc51fa549</uuid>
    <filterref filter = 'no-mac-spoofing'/>
    <filterref filter = 'no-ip-spoofing'/>
    <rule action = 'accept' direction = 'out' priority = '-650'>
      <mac protocolid = 'ipv4'/>
    </rule>
    <filterref filter = 'allow-incoming-ipv4'/>
    <filterref filter = 'no-arp-spoofing'/>
    <rule action = 'accept' direction = 'inout' priority = '-500'>
      <mac protocolid = 'arp'/>
    </rule>
    <filterref filter = 'no-other-l2-traffic'/>
    <filterref filter = 'qemu-announce-self'/>
  </filter>
```

第 1 行命令的输出是这个名为 clean-traffic 过滤器的 XML 定义。它引用了 no-mac-spoofing、no-ip-spoofing、allow-incoming-ipv4、no-arp-spoofing、no-other-l2-traffic、qemu-announce-self 等 6 个子过滤器，还有两个流量过滤规则。clean-traffic 过滤器的结构如图 5-46 所示。

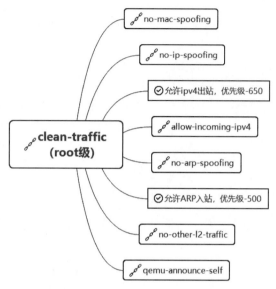

图 5-46　clean-traffic 过滤器结构

接下来执行的命令如下：

```
2 # virsh nwfilter-dumpxml no-mac-spoofing
   <filter name='no-mac-spoofing' chain='mac' priority='-800'>
     <uuid>5036c658-6bea-4299-aff4-0bd53a9048dd</uuid>
     <rule action='return' direction='out' priority='500'>
       <mac srcmacaddr='$MAC'/>
     </rule>
     <rule action='drop' direction='out' priority='500'>
       <mac/>
     </rule>
   </filter>

3 # virsh nwfilter-dumpxml no-ip-spoofing
   <filter name='no-ip-spoofing' chain='ipv4-ip' priority='-710'>
     <uuid>724ec182-c6fe-4289-a1d3-8b0b4821a897</uuid>
     <rule action='return' direction='out' priority='100'>
       <ip srcipaddr='0.0.0.0' protocol='udp'/>
     </rule>
     <rule action='return' direction='out' priority='500'>
       <ip srcipaddr='$IP'/>
     </rule>
```

```
    < rule action = 'drop' direction = 'out' priority = '1000'/>
</filter>
```

第 2 行命令输出了网络过滤器 no-mac-spoofing 的定义。第 3 行命令输出了网络过滤器 no-ip-spoofing 的定义。它们仅包含了流量过滤规则，而没有包含其他网络过滤器的引用。

提示：虚拟机的使用者有可能通过修改 IP 和 MAC 地址的方式来伪造主机身份，这给网络安全防护带来了巨大的挑战。可以通过配置网络过滤器 clean-traffic 来检查出站的流量中的 IP 及 MAC 地址。如果不是宿主机分配给虚拟机的 IP 和 MAC 地址，则认为是伪造的地址从而可以进行阻断。

5.11.4　网络过滤器语法基本格式

与 libvirt 管理的其他对象一样，网络过滤器也使用 XML 格式配置。从高层次看，格式如下：

```
< filter name = '过滤器名称' chain = '链的类型'>
  < uuid >链的 UUID </uuid>

  < rule ...>
     ....
  </rule>

  < filterref filter = 'XXXX'/>
</filter>
```

每个网络过滤器都有一个名称和 UUID，它们用作唯一标识符。网络过滤器可以具有 0 个或多个< rule >元素，它们用于定义网络控制。网络过滤器还可以通过 filterref 来引用其他的网络过滤器。

1. 链的类型

可以将过滤规则组织成链，从而形成树形结构。数据包从 root 链中开始进入评估流程，然后沿着分支链继续对其评估。既可以从这些分支链返回上一级链中，也可以被过滤规则直接丢弃或接收从而结束评估。

当激活网络过滤器时，libvirt 会自动为每个虚拟机的网络接口创建独立的网络过滤的 root 链。用户创建自定义的网络过滤器时，默认的类型是 root，也可以指定链的类型。

chain 是一个可选属性。使用它可以更好地组织过滤器，从而可以被防火墙子系统进行更有效的处理。它必须是以 mac、vlan、stp、arp、rarp、ipv4 或 ipv6 开头的。

示例 XML 文件如下：

```
# grep -h "chain" /etc/libvirt/nwfilter/*.xml
  <filter name='allow-arp' chain='arp' priority='-500'>
  <filter name='allow-dhcp-server' chain='ipv4' priority='-700'>
  <filter name='allow-dhcp' chain='ipv4' priority='-700'>
  <filter name='allow-incoming-ipv4' chain='ipv4' priority='-700'>
  <filter name='allow-ipv4' chain='ipv4' priority='-700'>
  <filter name='clean-traffic-gateway' chain='root'>
  <filter name='clean-traffic' chain='root'>
  <filter name='no-arp-ip-spoofing' chain='arp-ip' priority='-510'>
  <filter name='no-arp-mac-spoofing' chain='arp-mac' priority='-520'>
  <filter name='no-arp-spoofing' chain='root'>
  <filter name='no-ip-multicast' chain='ipv4' priority='-700'>
  <filter name='no-ip-spoofing' chain='ipv4-ip' priority='-710'>
  <filter name='no-mac-broadcast' chain='ipv4' priority='-700'>
  <filter name='no-mac-spoofing' chain='mac' priority='-800'>
  <filter name='no-other-l2-traffic' chain='root'>
  <filter name='no-other-rarp-traffic' chain='rarp' priority='-400'>
  <filter name='qemu-announce-self-rarp' chain='rarp' priority='-400'>
  <filter name='qemu-announce-self' chain='root'>
```

2．过滤规则

下面是一个简单的网络过滤器示例。它有这样一个规则，如果入站 IP 数据包中源 IP 地址是 192.168.122.21，则丢弃此数据包，示例代码如下：

```
<filter name='blockip' chain='ipv4'>
  <uuid>ebcdaeg1-a35f-62be-124e-208dadf2345a</uuid>
  <rule action='drop' direction='in' priority='500'>
    <ip match='yes' srcipaddr='192.168.122.21'/>
  </rule>
</filter>
```

rule 元素可包含如下属性。

（1）action：强制性的属性。必须是 drop、reject、accept、return 或 continue 中的一种。其中 drop 匹配规则无须进一步分析就静默丢弃数据包；reject 匹配规则将生成没有进一步分析结果的 ICMP 拒绝消息；accept 匹配规则接收无须进一步分析的数据包；return 匹配规则通过此过滤器，但将控制权返回给调用过滤器以进一步分析；continue 匹配规则继续到下一个规则以进一步分析。

（2）direction：强制性的属性。必须为 in、out 或者 inout 中的一种，分别针对入站、出站或出入站的流量。

（3）priority：可选的属性。它控制着规则相对于其他规则的实例化顺序。值较低的规则将在值较高的规则之前实例化。有效值的范围为 −1000～1000，默认值是 500。

（4）statematch：可选的属性。如果值是 0 或 false，则会关闭基于连接状态匹配。默认值为 true。

在本示例中＜ rule action＝'drop' direction＝'in' priority＝'500'＞，表示对入站的流量进行 drop 操作，其优先级为 500。

在 rule 中，需要设置与特定协议相关的匹配条件。一般模式是：

```
<协议名称 match = 'yes|no' attribute1 = 'value1' attribute2 = 'value2'/>
```

目前支持的协议包括 mac、arp、rarp、ip、ipv6、tcp/ip、icmp/ip、igmp/ip、udp/ip、udplite/ip、esp/ip、ah/ip、sctp/ip、tcp/ipv6、icmp/ipv6、igmp/ipv6、udp/ipv6、udplite/ipv6、esp/ipv6、ah/ipv6、sctp/ipv6。

match 的默认值是 yes。

在本示例中的＜ ip match＝'yes' srcipaddr＝'192.168.122.21'/＞，表示"如果源地址匹配 192.168.122.21"的过滤条件。

libvirt 所支持的协议及属性的表示方法，可参见 https://libvirt.org/formatnwfilter.html。

3. 变量的使用

为了提高网络过滤器的灵活性，可以在其中使用变量来替换固定的值。例如可以将上述示例中的网络过滤器进行修改，示例代码如下：

```
< filter name = 'blockip' chain = 'ipv4'>
  < uuid > ebcdaeg1 - a35f - 62be - 124e - 208dadf2345a </uuid>
  < rule action = 'drop' direction = 'out' priority = '500'>
    < ip match = 'no' srcipaddr = '$ IP'/>
  </rule>
</filter>
```

示例中的＜ ip match＝'no' srcipaddr＝'$ IP'/＞使用了一个名称为 IP 的变量。引用的变量始终要以 $ 符号为前缀。变量值的格式需要与属性的类型相匹配，例如 IP 参数必须包含点分十进制格式的 IP 地址。如果格式不正确，将会造成网络过滤器无法实例化，从而导致无法启动虚拟机或网络接口。

虚拟机 XML 文件引用这些过滤器的时候，可以提供变量名和值。下面的示例中，虚拟机在调用过滤器 blockip 时，提供了 3 次变量名 IP 和 IP 地址。libvirt 会创建 3 个单独的过滤规则，每个 IP 地址对应一个。这样会阻断来自这 3 个 IP 地址主机的访问。

示例代码如下：

```
   …
< devices >
  < interface type = 'bridge'>
    < mac address = '0a:1b:2c:3d:4e:5f'/>
```

```
      < filterref filter = 'blockip'>
        < parameter name = 'IP' value = '192.168.122.21'/>
        < parameter name = 'IP' value = '192.168.122.22'/>
        < parameter name = 'IP' value = '192.168.122.23'/>
      </filterref >
    </interface >
  </devices >
  ...
```

网络过滤器有一些保留的变量,其名称及含义如表 5-15 所示。

表 5-15 网络过滤器的保留变量

变量名称	含 义
MAC	接口的 MAC 地址
IP	接口使用的 IP 地址列表
IPV6	接口使用的 IPV6 地址列表
DHCPSERVER	受信任的 DHCP 服务器的 IP 地址列表
DHCPSERVERV6	受信任的 DHCP 服务器的 IPv6 地址列表
CTRL_IP_LEARNING	IP 地址检测方式的选择

最常用的保留变量是 MAC 和 IP,在预安装的网络过滤器中使用了它们。

变量 MAC 表示网络接口的 MAC 地址。如果未显式地提供,则 libvirt 守护程序会自动将其设置为网络接口的 MAC 地址。

变量 IP 表示虚拟机内部的操作系统应在给定接口上使用的 IP 地址。如果未显式地提供,则 libvirt 守护程序将尝试确定接口上正在使用的 IP 地址。

提示:如果网络过滤器中使用变量 IP 但未为其分配任何值,则将自动激活 libvirt 对虚拟机接口上使用的 IP 地址的检测。可以通过变量 CTRL_IP_LEARNING 来控制 IP 地址的学习方法,其有效值为 any、dhcp 或 none。默认值是 any,它表示 libvirt 可以使用任何数据包来确定虚拟机正在使用的地址。

5.11.5 自定义网络过滤器示例

下面我们通过示例来学习自定义过滤器的创建。假设现在要构建一个满足以下要求的网络过滤器:
(1) 防止虚拟机的 MAC、IP 和 ARP 欺骗。
(2) 仅打开虚拟机网络接口的 TCP 22 和 80 端口。
(3) 允许虚拟机向外发送 ping 流量,但不允许从外部 ping 虚拟机。
(4) 允许虚拟机执行 DNS 查询(UDP 53 的出站)。

示例 XML 文件如下：

```xml
<filter name='testfilter1'>
  <!-- 引用clean-traffic以防止MAC、IP和ARP欺骗.不提供MAC和IP地址,由libvirt自动探查出VM的实际值 -->
  <filterref filter='clean-traffic'/>

  <!-- 允许TCP端口22(ssh)的入站 -->
  <rule action='accept' direction='in'>
    <tcp dstportstart='22'/>
  </rule>

  <!-- 允许TCP端口80(http)的入站 -->
  <rule action='accept' direction='in'>
    <tcp dstportstart='80'/>
  </rule>

  <!-- 允许出站的ICMP echo请求 -->
  <rule action='accept' direction='out'>
    <icmp type='8'/>
  </rule>

  <!-- 允许入站的ICMP echo响应请求 -->
  <rule action='accept' direction='in'>
    <icmp type='0'/>
  </rule>

  <!-- 允许出站的UDP DNS查找 -->
  <rule action='accept' direction='out'>
    <udp dstportstart='53'/>
  </rule>

  <!-- 丢弃其他所有流量 -->
  <rule action='drop' direction='inout'>
    <all/>
  </rule>
</filter>
```

预安装的 clean-traffic 网络过滤器已满足防止欺骗的要求，因此可以直接引用它，然后通过前两条规则分别允许 TCP 端口 22 和 80 的入站，接下来的 2 条规则用于控制 ping，随后是允许 DNS 查询。为了禁止所有其他流量出入站，最后添加一条丢弃所有其他流量的规则。

这个过滤器被命令为 testfilter1，在虚拟机的网络接口 XML 中可以这样来引用它，代码如下：

```
[ … ]
    < interface type = 'network'> < mac address = '52:54:00:27:5f:c9'/>
    < source network = 'default'/>
    < model type = 'virtio'/>
    < filterref filter = 'testfilter1'/>
    < address type = 'pci' domain = '0x0000' bus = '0x00' slot = '0x03' function = '0x0'/>
  </interface >
[ … ]
```

5.12 本章小结

本章讲解了虚拟网络的管理，包括隔离、NAT、桥接、路由、开放等网络类型的原理与配置，以及 VLAN 和网络过滤器的原理与配置。

第 6 章将讲解管理虚拟存储。

第 6 章 管理虚拟存储

虚拟存储是可以分配给虚拟机的存储空间,是宿主机物理存储的一部分,是通过模拟或半虚拟化的块设备驱动程序分配给虚拟机的存储。

与虚拟网络类似,我们需要一个存储后端为虚拟机提供存储。如果是由 libvirt 创建和管理的存储,则被称为托管的存储(Managed Storage),它使用存储池、存储卷等技术灵活高效地管理存储。如果是由管理员"手工"创建和管理的存储,则被称为非托管的存储(Unmanaged Storage),它适用于小型环境、测试环境或排错。

本章要点:
- 存储原理。
- 非托管的存储。
- qemu-img 命令的使用。
- 存储池。
- 存储卷。

6.1 虚拟存储的术语

6.1.1 虚拟机的存储设备

宿主机可以为虚拟机提供多种类型的存储设备。每种类型的存储设备适合于不同的场景,例如:最小空间使用、高 I/O 性能、操作系统兼容性等。

virsh 有多个以 domblk 开头的子命令,它们可提供虚拟机块存储设备的信息。

(1) domblkerror:显示块设备上的错误。

(2) domblkinfo:显示块设备的基本信息。

(3) domblklist:显示所有块设备的列表。

(4) domblkstat:显示块设备的状态。

提示:dom 是 domain 的缩写,在 KVM 中,domain 与 guest、virtual machine 含义相同。blk 是 block 的缩写。

执行的命令如下：

```
1 # virsh domblklist centos6.10
  Target     Source
  ------------------------------------------------
  vda        /var/lib/libvirt/images/centos6.10.qcow2
  hda        -

2 # virsh domblkinfo centos6.10 vda
  Capacity:      9663676416
  Allocation:    1442623488
  Physical:      9888071680

3 # virsh domblkstat centos6.10
   rd_req 24
   rd_Bytes 700416
   wr_req 0
   wr_Bytes 0
   flush_operations 0
   rd_total_times 22826241
   wr_total_times 0
   flush_total_times 0
```

第 1 行命令会输出虚拟机 centos6.10 的块设备的列表。有 2 个存储目标（Target），vda 的源是一个虚拟磁盘文件（/var/lib/libvirt/images/centos6.10.qcow2），had（CDROM）当前没有介质文件。

第 2 行命令显示了虚拟磁盘 vda 的容量信息。

第 3 行命令显示了虚拟磁盘的读写信息的统计结果。

接下来执行的命令如下：

```
4 # virsh dumpxml centos6.10
    ...
    <devices>
      <emulator>/usr/libexec/qemu-kvm</emulator>
      <disk type='file' device='disk'>
        <driver name='qemu' type='qcow2'/>
        <source file='/var/lib/libvirt/images/centos6.10.qcow2'/>
        <target dev='vda' bus='virtio'/>
        <address type='pci' domain='0x0000' bus='0x00' slot='0x07' function='0x0'/>
      </disk>
      <disk type='file' device='cdrom'>
        <driver name='qemu' type='raw'/>
        <target dev='hda' bus='ide'/>
        <readonly/>
```

```
            <address type='drive' controller='0' bus='0' target='0' unit='0'/>
         </disk>
         ...
```

第 4 行命令显示了虚拟机两个块设备的详细配置。可以对照 virt-manager 中虚拟磁盘的信息来看,如图 6-1 所示。

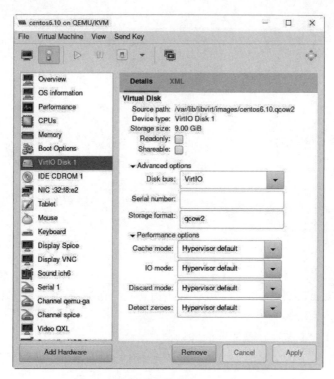

图 6-1　虚拟磁盘信息(virt-manager)

1) <disk type='file' device='disk'>

type:可能的值有 file、block、dir、network、volume 或 nvme。本示例是 file,它表示映像文件。

device:可能的值有 floppy、disk、cdrom 或 lun,默认值为 disk。

2) <driver name='qemu' type='qcow2'/>

指定驱动程序的配置参数。Hypervisor 可支持多种后端驱动程序,所以 name 属性用于设置驱动程序名称,例如 qemu、xen 等。如果 name='xen',则表示支持的类型有 tap、tap2、phy、file 和 aio。如果 name='qemu',则表示支持的类型有 raw、qcow、qcow2 和 qed。

3) <source file='/var/lib/libvirt/images/centos6.10.qcow2'/>

由于类型是 qcow2,所以 source 指定了一个 qcow2 类型文件的位置。

4) < target dev = 'vda' bus = 'virtio'/>

target 控制着提供给虚拟机操作系统的总线/设备。dev 属性仅仅是一个逻辑设备名称,并不能保证在虚拟机操作系统中看到的就是这个名称。bus 属性用于指定要模拟的主线的类型,可能的值有 ide、scsi、virtio、xen、usb、sata 和 sd。

在 virt-manager 中,还可以设置一些额外的属性,包括以下几种属性:

1) readonly

如果被选中,则虚拟机无法向该设备写入数据。当 device = 'cdrom'时,这是默认设置。

2) shareable

如果被选中,则表明该设备可以在虚拟机之间共享(需要 Hypervisor 和虚拟机操作系统支持此功能)。此选项会自动禁用该设备的缓存。

3) Cache mode

控制缓存的工作机制。可能的值有 default、none、writethrough、writeback、directsync(类似 writethrough,但绕过了宿主机页面缓存)和 unsafe(宿主机可能会缓存所有磁盘 IO,而虚拟机的同步请求将被忽略)。

4) I/O Mode

控制有关 I/O 的策略。可能的值有 threads 和 native。

5) Discard Mode

丢弃请求也称为修剪(trim)或取消映射(unmap)。它控制丢弃请求是被忽略或还是被传递给文件系统。可能的值有 unmap(允许传递丢弃请求)和 ignore(忽略丢弃请求)。

6) Detect zeroes

控制是否检测零写入请求。可能的值有 on、off 和 unmap。前两个值分别用于打开和关闭检测,第 3 个值 unmap 开启检测功能,并尝试根据上述 Discard Mode 的值从映射中丢弃这些区域(如果将 Discard Mode 设置为 ignore,则表示它被当作 on)。注意:打开检测功能是一项计算密集型操作,但可以节省访问慢速介质的访问时间和空间。

提示:通常无须设置 Performance options 的选项,因为保持默认的值(Hypervisor default)即可满足大多数需求。

Cockpit、virt-manager 并没有提供所有的选项参数,只允许使用一些常用的选项参数。例如在 virt-manager 中可以设置的总线类型有 IDE、sata、scsi、usb 和 virtio,而在 Cockpit 中仅有 sata、scsi、usb 和 virtio 等 4 种类型,如图 6-2 所示。

6.1.2 宿主机的存储资源

可以分配给虚拟机的存储资源有很多种,常用的有映像文件、LVM 卷、宿主机设备和分布式存储系统。

1. 映像文件

映像文件只能存储在宿主机文件系统上,例如本地文件系统或网络文件系统(NFS 或 CIFS)。

图 6-2　虚拟磁盘信息（Cockpit）

使用 dd、qemu-img 工具可以创建映像文件，使用 qemu-img、libguestfs 等工具可以管理、备份和监视映像文件。libvirt 的存储卷管理功能可以满足大多数映像文件的管理需求。

执行的命令如下：

```
1 # qemu-img --version
  qemu-img version 2.12.0 (qemu-kvm-2.12.0-99.module_el8.2.0+524+f765f7e0.4)
  Copyright (c) 2003-2017 Fabrice Bellard and the QEMU Project developers

2 # qemu-img --help | grep Supported
  Supported formats: blkdebug blkreplay blkverify copy-on-read file ftp ftps gluster host_cdrom host_device http https iscsi iser luks nbd null-aio null-co nvme qcow2 quorum raw rbd ssh throttle vhdx vmdk vpc
```

映像文件有很多种格式，第 2 行命令列出了 v2.12.0 版本的 qemu-img 所支持的格式。KVM 最常用的磁盘映像格式是 raw 和 qcow2。

raw 格式的映像文件只包含数据内容，而没有其他元数据。如果文件系统支持，则 raw 格式文件既可以是预分配的（pre-allocated），也可以是稀疏的（sparse）。稀疏文件可以按需分配宿主机磁盘空间，因此它是精简配置（thin provisioning）的一种形式。预分配的文件中有可能会有长期"闲置"的空间，但它比稀疏文件具有更高的性能。当对磁盘 I/O 性能要求高且很少需要通过网络传输映像文件时，建议使用 raw 文件。

qcow2 映像文件提供了许多高级功能，包括派生、快照、压缩和加密等。因为在映像文件中仅分配了数据的扇区，所以 qcow2 文件可以更有效地通过网络进行传输。

2. LVM 卷

可以将 LVM 的逻辑卷直接分配给虚拟机，所以它提供了比映像文件更高的性能。LVM 的精简供给、快照功能可以更加高效地使用存储空间。由于还可以用宿主机上的 LVM 工具进行管理，所以管理更简单。

3. 宿主机设备

可以将宿主机上的设备(例如物理 CD-ROM、磁盘、分区和 LUN)"原封不动"地分配给虚拟机,从而获得"原生"的性能和特性。

4. 分布式存储系统

分布式存储系统是新型的存储解决方案,它是将数据分散存储在多台独立的设备上,各司其职、协同合作,统一地对外提供存储服务。这种可扩展的系统结构,不但提高了系统的可靠性、可用性和存取效率,还易于扩展。KVM 目前支持 Gluster、Ceph 等多种分布式存储系统。

6.2 非托管的存储

与虚拟机一起使用但不被 libvirt 直接控制和监视的存储,被称为非托管存储(Unmanaged Storage)。只要拥有足够权限,就可以直接将宿主机上的映像文件或设备分配给虚拟机。这是最快捷的将额外存储添加给虚拟机的方法。

非托管存储的最大缺点是不灵活,它们在虚拟机上的配置文件的路径是固定的、硬编码的,一旦虚拟磁盘文件位置发生变化,就需要修改虚拟机的配置文件。这有些像通过 IP 地址访问主机,如果 IP 地址发生变化就比较麻烦。如果使用了主机的别名或域名这个"中间层",就会很灵活。托管的存储引用了"存储池"这一概念,它就是低层的存储与上层的虚拟机之间的"中间层",所以更加灵活。

非托管存储目前主要用于小型环境、测试环境或排错。对于初学者来讲,先掌握一些非托管存储知识,然后学习托管的存储会比较轻松一些。

下面用映像文件来做几个非托管存储的实验。

6.2.1 使用 dd 创建磁盘映像文件

磁盘映像文件是存储在宿主机文件系统中的普通文件。在 RHEL/CentOS 8 中创建磁盘映像文件常用的工具是 dd 和 qemu-img。

dd 是一个 UNIX 和类 UNIX 系统上的命令,其主要功能是转换和复制文件。下面,我们就使用它来创建 raw 格式的文件,命令如下:

```
1 #dd if = /dev/zero of = /vm/test1.img bs = 1M count = 1024
  1024 + 0 records in
  1024 + 0 records out
  1073741824 Bytes (1.1 GB, 1.0 GiB) copied, 0.290342 s, 3.7 GB/s

2 #ls - lhs /vm/test1.img
  1.0G - rw - r - - r - - . 1 root root 1.0G Dec 19 15:22 /vm/test1.img

3 #du - sh /vm/test1.img
  1.0G/vm/test1.img
```

第 1 行命令将使用 1MB 大小的块（1bs＝1MB），将来自输入文件（if）/dev/zero（无限提供 0 的设备）数据复制到输出文件（of）/vm/test1.img（磁盘映像）中，这个复制操作重复 1024 次（count＝1024）。

第 2 行命令显示了新的文件信息。其中-s 选项是以块为单位显示每个文件的分配大小。新文件大小是 1GB，即使现在其中没有任何数据，其空间占用也是 1GB。

第 3 行的 du 命令的输出也显示了此文件空间占用是 1GB。

接下来执行的命令如下：

```
4 # stat /vm/test1.img
    File: /vm/test1.img
    Size: 1073741824    Blocks: 2097152    IO Block: 4096    regular file
Device: fd02h/64770d   Inode: 151         Links: 1
Access: (0644/-rw-r--r--)   Uid: (    0/   root)   Gid: (    0/   root)
Context: unconfined_u:object_r:unlabeled_t:s0
Access: 2020-12-19 15:22:36.295746362 +0800
Modify: 2020-12-19 15:22:36.584746361 +0800
Change: 2020-12-19 15:22:36.584746361 +0800
 Birth: -
```

第 4 行 stat 命令会输出这个文件更详细的信息。这个文件共使用了 2097152 个块。

接下来执行的命令如下：

```
5 # file /vm/test1.img
  /vm/test1.img: data

6 # qemu-img info /vm/test1.img
  image: /vm/test1.img
  file format: raw
  virtual size: 1.0G (1073741824 Bytes)
  disk size: 1.0G
```

第 5 行 file 命令显示这个文件的类型是数据型，而第 6 行 qemu-img 的 info 子命令则会显示更详细的信息，包括文件格式、虚拟磁盘文件大小、真实空间占用等。

第 1 行命令生成文件的方法是预分配，即使没有数据也会分配磁盘空间。如果希望减少这个映像文件的大小，就需要使用精简配置技术。

在传统的计算机文件中，"空"的部分还会占用存储空间。为了解决这一问题，就诞生了稀疏文件（Sparse File）。它不保存空字节，而是使用空间占用很少的元数据（metadata）表达它们，这会更有效地利用文件系统空间。仅当块中包含"真实"（非空）的数据时，整个块大小才作为实际大小写入磁盘。在读取稀疏文件时，文件系统将用元数据表示的空块转换为填充有空字节的"实际"的块。这些特性对应用程序是透明的，它们不知道这种转换。现在很多文件系统都支持稀疏文件，包括 ext2、ext3、ext4、xfs、ntfs 等。

预分配和精简配置(稀疏)是空间分配的方法,或者可以将其称为格式,其各有优缺点。

(1) 预分配:创建预分配的映像文件时会立即分配空间。向预分配格式映像文件写入的速度要比精简配置格式快一些。

(2) 精简配置:根据实际需要分配空间。例如,如果创建一个稀疏分配的映像文件,则最初占用的空间很少,随着数据的写入而逐渐增长,直到达到文件的最大值。可以使用精简配置实现存储的超量分配。

在 dd 命令中使用 seek 选项,可以创建精简配置的映像文件,命令如下:

```
7 #dd if = /dev/zero of = /vm/test2 - seek.img bs = 1M seek = 1024 count = 0
  0 + 0 records in
  0 + 0 records out
  0 Bytes copied, 5.3893e - 05 s, 0.0 kB/s

8 #ls - lsh /vm/test2 - seek.img
  0 - rw - r - - r - - . 1 root root 1.0G Dec 19 15:24 /vm/test2 - seek.img

9 #du - sh /vm/test2 - seek.img
  0     /vm/test2 - seek.img
```

第 7 行的 dd 命令使用了 seek=1024 选项,则会在输出文件开头先跳过 1024 个块后再开始复制,这就生成了一个 1MB×1024=1024MB(1GB)大小的"空洞"。由于使用了 count=0 选项,所以 dd 命令并没有复制数据,这样生成的新文件就是一个完全"空"的文件。

第 8 行命令显示了这个新文件的大小是 1GB,但是空间占用却是 0。这个结论也可以在第 9 行命令的输出中得到验证。

接下来执行的命令如下:

```
10 #stat /vm/test2 - seek.img
     File: /vm/test2 - seek.img
     Size: 1073741824    Blocks: 0     IO Block: 4096    regular file
   Device: fd02h/64770d    Inode: 152    Links: 1
   Access: (0644/ - rw - r - - r - - )  Uid: (  0/  root)  Gid: (  0/  root)
   Context: unconfined_u:object_r:unlabeled_t:s0
   Access: 2020 - 12 - 19 15:24:42.587745895 + 0800
   Modify: 2020 - 12 - 19 15:24:42.587745895 + 0800
   Change: 2020 - 12 - 19 15:24:42.587745895 + 0800
    Birth: -

11 #qemu - img info /vm/test2 - seek.img
   image: /vm/test2 - seek.img
   file format: raw
   virtual size: 1.0G (1073741824 Bytes)
   disk size: 0
```

第 10 行 stat 命令会输出这个文件更详细的信息,这个文件使用的块数是 0。

第 11 行命令显示了该文件真实空间占用也是 0。

6.2.2 使用 virsh 管理虚拟机磁盘映像文件

可以使用 virsh 的 attach-disk 和 attach-device 子命令将映像文件附加给虚拟机。attach-disk 使用命令行参数进行配置,attach-device 使用 XML 文件进行配置。

attach-disk 子命令如下:

```
1 # virsh attach-disk --help
    NAME
        attach-disk - attach disk device

    SYNOPSIS
        attach-disk <domain> <source> <target> [--targetbus <string>] [--driver <string>]
[--subdriver <string>] [--iothread <string>] [--cache <string>] [--io <string>] [--type
<string>] [--mode <string>] [--sourcetype <string>] [--serial <string>] [--wwn
<string>] [--rawio] [--address <string>] [--multifunction] [--print-xml] [--
persistent] [--config] [--live] [--current]

    DESCRIPTION
        Attach new disk device.

    OPTIONS
        [--domain] <string>     domain name, id or uuid
        [--source] <string>     source of disk device
        [--target] <string>     target of disk device
        --targetbus <string>    target bus of disk device
        --driver <string>       driver of disk device
        --subdriver <string>    subdriver of disk device
        --iothread <string>     IOThread to be used by supported device
        --cache <string>        cache mode of disk device
        --io <string>           io policy of disk device
        --type <string>         target device type
        --mode <string>         mode of device reading and writing
        --sourcetype <string>   type of source (block|file)
        --serial <string>       serial of disk device
        --wwn <string>          wwn of disk device
        --rawio                 needs rawio capability
        --address <string>      address of disk device
        --multifunction         use multifunction pci under specified address
        --print-xml             print XML document rather than attach the disk
        --persistent            make live change persistent
        --config                affect next boot
        --live                  affect running domain
        --current               affect current domain
```

(1) [--domain] <string>:虚拟机的标识,可以是名称、ID 或 UUID。

(2) [--source] < string >：源磁盘设备，在本示例中是映像文件的路径名。

(3) [--target] < string >：目标磁盘设备名称。

(4) --config：指定下次启动后生效。

(5) --live：修改正在运行的虚拟机。

(6) --current：根据虚拟机的状态而定，如果虚拟机正在运行，就修改当前的状态，相当于--live；如果虚拟机没有运行，就修改配置文件，相当于--config。

(7) --driver：对于 QEMU 来讲是 qemu。对于 Xen 来讲可以是 file、tap 或 phy。

(8) --subdriver：向驱动程序传递更多详细信息。对于 QEMU，subdriver 应匹配映像文件的格式，例如 raw 或 qcow2（默认为 raw 格式）。对于 Xen，subdriver 可以是 aio。

接下来执行的命令如下：

```
2 # virsh domblklist centos6.10
  Target     Source
  ------------------------------------------------
  vda        /var/lib/libvirt/images/centos6.10.qcow2
  hda        -

3 # virsh attach-disk centos6.10 /vm/test1.img vdb --config --live
  error: Failed to attach disk
  error: Requested operation is not valid: domain is not running

4 # virsh attach-disk centos6.10 /vm/test1.img vdb --config
  Disk attached successfully
```

第 2 行命令用于查看虚拟机的块设备列表。

第 3 行命令将 /vm/test1.img 附加到虚拟机的 vdb 接口。由于当前虚拟机并没有运行，所以使用--live 选项会出现错误提示。

第 4 行命令中没有--live 选项，所以附加成功。

接下来执行的命令如下：

```
5 # virsh start centos6.10

6 # virsh attach-disk centos6.10 /vm/test2-seek.img vdc --config --live
  Disk attached successfully

7 # virsh domblklist centos6.10
  Target     Source
  ------------------------------------------------
  vda        /var/lib/libvirt/images/centos6.10.qcow2
  vdb        /vm/test1.img
  vdc        /vm/test2-seek.img
  hda        -
```

第 5 行命令启动虚拟机后,第 6 行命令使用--config 和--live 选项,这既修改了虚拟机的配置文件也更新了当前的状态,成功地将/vm/test2-seek.img 附加到虚拟机的 vdc 接口。

第 7 行命令再次查看虚拟机当前的块设备列表。

接下来执行的命令如下：

```
8 # virsh dumpxml centos6.10
  ...
  <devices>
    <emulator>/usr/libexec/qemu-kvm</emulator>
    <disk type='file' device='disk'>
      <driver name='qemu' type='qcow2'/>
      <source file='/var/lib/libvirt/images/centos6.10.qcow2'/>
      <backingStore/>
      <target dev='vda' bus='virtio'/>
      <alias name='virtio-disk0'/>
      <address type='pci' domain='0x0000' bus='0x00' slot='0x07' function='0x0'/>
    </disk>
    <disk type='file' device='disk'>
      <driver name='qemu' type='raw'/>
      <source file='/vm/test1.img'/>
      <backingStore/>
      <target dev='vdb' bus='virtio'/>
      <alias name='virtio-disk1'/>
      <address type='pci' domain='0x0000' bus='0x00' slot='0x09' function='0x0'/>
    </disk>
    <disk type='file' device='disk'>
      <driver name='qemu' type='raw'/>
      <source file='/vm/test2-seek.img'/>
      <backingStore/>
      <target dev='vdc' bus='virtio'/>
      <alias name='virtio-disk2'/>
      <address type='pci' domain='0x0000' bus='0x00' slot='0x0a' function='0x0'/>
    </disk>
    <disk type='file' device='cdrom'>
      <driver name='qemu'/>
      <target dev='hda' bus='ide'/>
      <readonly/>
      <alias name='ide0-0-0'/>
      <address type='drive' controller='0' bus='0' target='0' unit='0'/>
    </disk>
  ...
```

第 8 行命令输出了虚拟机的详细配置,可以看到 3 个块设备的配置。从< driver name='qemu' type='raw'/>可以看出,默认驱动是 qemu、虚拟磁盘类型是 raw。如果后续使用 qcow2 的格式,就必须使用--subdriver qcow2 选项来指定映像文件格式。

由于此时虚拟机正在运行,所以会发现有两个新的元素 backingStore 和 alias。backingStore 指定 source 文件的基础映像文件,目前为空。alias 指定设备的别名。

与 attach-disk 相对应,virsh 的 detach-disk 子命令可以将存储设备从虚拟机上分离,命令如下:

```
9 # virsh detach-disk --help
  NAME
    detach-disk - detach disk device

  SYNOPSIS
    detach-disk <domain> <target> [--persistent] [--config] [--live]
    [--current] [--print-xml]

  DESCRIPTION
    Detach disk device.

  OPTIONS
    [--domain] <string>  domain name, id or uuid
    [--target] <string>  target of disk device
    --persistent         make live change persistent
    --config             affect next boot
    --live               affect running domain
    --current            affect current domain
    --print-xml          print XML document rather than detach the disk
```

第 9 行命令显示了 detach-disk 子命令的帮助信息,相对于 attach-disk,它要简单很多。接下来执行的命令如下:

```
10 # virsh detach-disk centos6.10 vdb --config --live
   Disk detached successfully

11 # virsh detach-disk centos6.10 vdc --config --live
   Disk detached successfully

12 # virsh domblklist centos6.10 --inactive
   Target   Source
   ------------------------------------------------
   vda      /var/lib/libvirt/images/centos6.10.qcow2
   hda      -
```

第 10 行、第 11 行命令将 vdb 和 vdc 从虚拟机上成功分离,第 12 命令用于检查确认。

提示:在执行 virsh detach-disk 之前,要保证虚拟机中没有应用程序还在使用虚拟磁盘上的数据,否则有可能造成数据丢失、文件损坏。

6.2.3 使用 virt-manager 管理虚拟机磁盘映像文件

virt-manager 可以添加非托管的存储资源，但是添加成功之后，还会为其创建新的存储池，从而变成了托管的存储。下面通过实验来验证一下。

首先创建全新的目录及新的映像文件，命令如下：

```
#mkdir /vm2
#dd if=/dev/zero of=/vm2/test3-seek.img bs=1M seek=1024 count=0
0+0 records in
0+0 records out
0 Bytes copied, 6.8644e-05 s, 0.0 kB/s
```

然后在 virt-manager 中进行如下操作。

（1）在 virt-manager 中选中虚拟机 centos6.10。单击 按钮，会出现硬件的详细信息窗口，如图 6-3 所示。

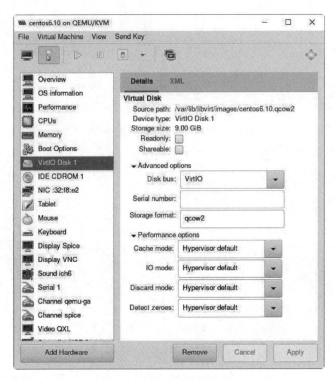

图 6-3 硬件的详细信息窗口

（2）单击 Add Hardware 按钮，会出现 Add New Virtual Hardware 窗口。确保在左框"硬件类型"窗格中选中了 Storage，如图 6-4 所示。

（3）选中 Select or create custom storage，然后单击 Manage 按钮。

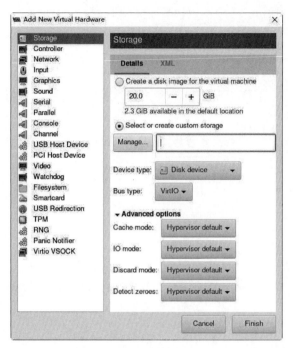

图 6-4 添加新虚拟硬件窗口

(4) 在 Choose Storage Volume 对话窗口中单击 Browse Local 按钮,如图 6-5 所示。

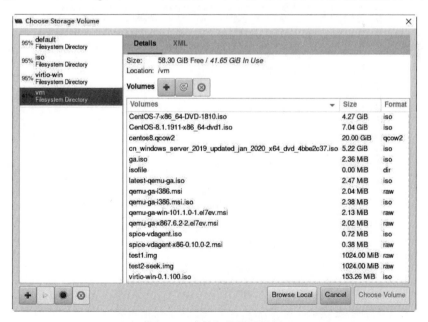

图 6-5 选择存储卷对话框

（5）在 Locate existing storage 对话窗口中找到/vm2/test3-seek.img 文件。单击 Open 按钮，如图 6-6 所示。

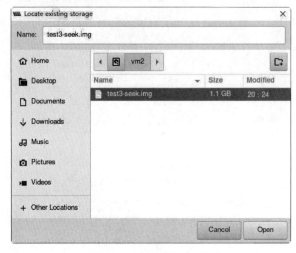

图 6-6　查找现有存储

（6）返回 Add New Virtual Hardware 窗口后，单击 Finish 按钮完成添加，如图 6-7 所示。

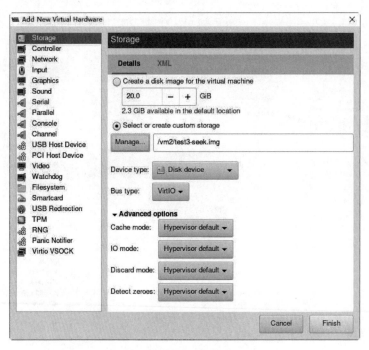

图 6-7　添加新虚拟硬件窗口

virt-manager 自动将 /vm2 目录创建为一个基于目录的存储池,而 /vm2/test3-seek.img 则是这个存储池中的一个存储卷。

(7) 打开 virt-manager 的 Edit 菜单,单击 Connection Details。在连接细节窗口中单击 Storage 选项卡,就可以看到存储池 vm2 和存储卷 test3-seek.img 了,如图 6-8 所示。

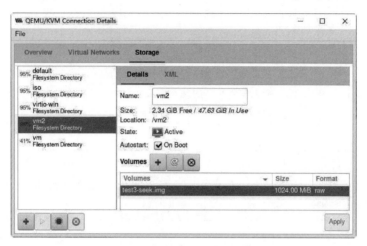

图 6-8 新的存储池和存储卷

我们也可以通过 virsh 的 pool-list 和 vol-list 来查看新的存储池和存储卷,命令如下:

```
# virsh pool-list
 Name                 State      Autostart
-------------------------------------------
 default              active     yes
 iso                  active     yes
 virtio-win           active     yes
 vm                   active     yes
 vm2                  active     yes

# virsh vol-list vm2
 Name                 Path
-------------------------------------------
 test3-seek.img       /vm2/test3-seek.img
```

提示:当前版本 Cockpit(cockpit-machines-211.3)还不支持非托管存储的使用。

6.3 qemu-img 命令的使用

qemu-img 是一个管理虚拟磁盘映像文件的命令行工具。它可以创建、格式化、修改、验证与维护多种格式的映像文件。常见的子命令如表 6-1 所示。

表 6-1 qemu-img 常用子命令

子命令	描述
create	创建新的映像文件
check	检查现有映像文件是否有错误
convert	转换映像文件格式
info	显示映像文件的信息
snapshot	管理映像文件的快照
commit	将保存在派生映像文件的数据提交到基础映像文件中
rebase	调整派生映像文件的基础映像文件的位置与名称
resize	调整映像文件的大小

警告：切勿使用 qemu-img 修改正在运行的虚拟机或任何其他进程正在使用的映像，这可能会破坏映像文件。另外，查询正在被另外一个进程使用的映像文件（例如：虚拟机正在运行），也有可能会获得不准确的信息。

6.3.1 qemu-img 支持的映像文件格式

使用 qemu-img 命令时，常常需要使用-f 或-F 来指定映像文件格式。当前版本支持以下格式。

（1）raw：原始磁盘映像格式（默认）。这是最快的映像文件的格式。如果文件系统支持稀疏文件（例如 ext2、ext3、ext4 或 xfs），则只有在写入数据时才会分配空间。可以使用 ls -ls、qemu-img info 命令获得映像文件使用的实际大小。尽管 raw 格式映像文件可提供最佳性能，但是它支持的功能有限，例如：不支持快照功能。

（2）qcow2：QEMU 映像文件格式，是功能丰富的流行格式。支持 AES 加密，基于 zlib 的压缩，支持快照，体积较小。不过，这些功能是以性能开销为代价的。

（3）bochs：Bochs 磁盘映像格式。

（4）cloop：Linux 压缩循环映像。

（5）cow：用户模式下的 Copy On Write 映像格式。支持该格式是为了与以前的版本兼容。

（6）dmg：Mac 磁盘映像格式。

（7）nbd：网络块设备。

（8）parallels：并行虚拟化磁盘映像格式。

（9）qcow：早期的 QEMU 映像格式。支持该格式是为了与以前的版本兼容。

（10）qed：早期的 QEMU 映像格式。支持该格式是为了与以前的版本兼容。

（11）vdi：Oracle VM VirtualBox 虚拟磁盘映像格式。

（12）vhdx：Microsoft Hyper-V 虚拟磁盘映像格式。

(13) vmdk：VMware 虚拟磁盘映像格式。

(14) vvfat：虚拟 VFAT 磁盘映像格式。

6.3.2　创建和格式化新的映像文件

创建新的磁盘映像文件语法格式如下：

```
#qemu-img create [-f format] [-o options] filename [size]
```

可以通过-f 选项指定格式，默认格式是 raw 格式，命令如下：

```
1 #qemu-img create test1.img 512M
  Formatting 'test1.img', fmt=raw size=536870912

2 #qemu-img info test1.img
  image: test1.img
  file format: raw
  virtual size: 512M (536870912 Bytes)
  disk size: 4.0K
```

从第 2 行命令的输出可以看出：默认格式是 raw 格式，而且是精简供给（稀疏）格式。接下来执行的格式如下：

```
3 #qemu-img create -f qcow2 test2.qcow2 512M
  Formatting 'test2.qcow2', fmt=qcow2 size=536870912 cluster_size=65536 lazy_refcounts=off refcount_bits=16

4 #qemu-img info test2.qcow2
  image: test2.qcow2
  file format: qcow2
  virtual size: 512M (536870912 Bytes)
  disk size: 196K
  cluster_size: 65536
  Format specific information:
      compat: 1.1
      lazy refcounts: false
      refcount bits: 16
      corrupt: false

5 #ls -lhs test*
  4.0K -rw-r--r--. 1 root root 512M Dec 20 22:28 test1.img
  196K -rw-r--r--. 1 root root 193K Dec 20 22:29 test2.qcow2
```

第 3 行命令使用-f qcow2 选项指定了新映像文件的格式。创建 qcow2 格式的文件的时候,会按默认选项进行格式化。可以将 qcow2 映像文件格式简单理解成由三部分组成:元数据(Metadata)、簇的位图(Cluster Data)和簇的数据(Cluster Data),如图 6-9 所示。

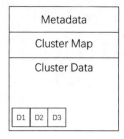

图 6-9　qcow2 文件格式的简化示意图

从第 4 行命令的输出可以看出:qcow2 格式的映像文件属性更丰富一些。

第 5 行命令列出了两个文件的信息。在操作系统中显示 qcow2 格式的文件大小是 193KB,它并不能真实地反映其可以存储数据的空间大小(512MB)。

默认的 raw 类型是精简供给(稀疏)格式。可以通过 preallocation 选项来调整格式,它允许的值包括以下几个。

(1) off:默认值。不进行预分配空间,即稀疏格式。

(2) falloc:调用 posix_fallocate()函数来预分配空间。

(3) full:通过将零写入底层存储来预分配空间,类似 dd if=/dev/zero。

接下来执行的命令如下:

```
6 # qemu-img create -f raw -o preallocation=full test.raw 50M
  Formatting 'test.raw', fmt=raw size=52428800 preallocation=full

7 # qemu-img info test.raw
  image: test.raw
  file format: raw
  virtual size: 50M (52428800 Bytes)
  disk size: 50M
```

提示:falloc 是 fast allocation 的缩写。posix_fallocate()函数分配文件块的方法是:标记使用但不进行未初始化,所以创建映像文件的速度比 full 模式快很多,这类似于 VMware ESXi 的"厚置备延迟置零"。

6.3.3　检查映像文件的一致性

对 qcow2、vdi、vhdx、vmdk 和 qed 等格式的映像文件执行一致性检查,其语法格式如下:

```
# qemu-img check [-f format] imgname
```

示例命令如下:

```
1 # qemu-img check test1.img
  qemu-img: This image format does not support checks

2 # qemu-img check test2.qcow2
  No errors were found on the image.
  Image end offset: 262144
```

第 1 行命令的输出显示了 qemu-img 不支持对 raw 格式的映像文件进行检查。

如果映像文件没有问题,则会输出 No errors were found on the image 提示信息。如果指定了"-r"选项,则 qemu-img 会尝试修复在检查过程中发现的所有不一致之处。

6.3.4 重新调整映像文件的大小

qemu-img 可以增加或收缩 raw 格式的映像文件大小,但只能增加 qcow2 映像文件的大小而不能收缩。

重新调整映像文件大小的语法格式如下:

```
# qemu-img resize filename size
```

示例命令如下:

```
1 # qemu-img resize test1.img 1G
  WARNING: Image format was not specified for 'test1.img' and probing guessed raw.
  Automatically detecting the format is dangerous for raw images, write operations on block 0
  will be restricted.
  Specify the 'raw' format explicitly to remove the restrictions.
  Image resized.

2 # qemu-img info test1.img
  image: test1.img
  file format: raw
  virtual size: 1.0G (1073741824 Bytes)
  disk size: 4.0K
```

在调整大小时,建议通过-f 选项明确指明文件的格式。如果未指明,则 qemu-img 会自动进行判断,这对于 raw 格式映像文件是有风险的,所以第 1 行命令的输出中有警告信息。

另外,还可以使用相对值来调整大小,其语法格式如下:

```
# qemu-img resize filename [+|-]size[K|M|G|T]
```

接下来执行的命令如下：

```
3 # qemu-img resize test2.qcow2 +512M
  Image resized.

4 # qemu-img info test2.qcow2
  image: test2.qcow2
  file format: qcow2
  virtual size: 1.0G (1073741824 Bytes)
  disk size: 200K
  cluster_size: 65536
  Format specific information:
      compat: 1.1
      lazy refcounts: false
      refcount bits: 16
      corrupt: false
```

> **警告**：在收缩映像文件之前，除了需要做好备份之外，还必须在虚拟机中使用工具来减少文件系统和分区大小，否则有可能会造成数据丢失。

6.3.5　qcow2映像文件的选项

qcow2是常用的映像文件格式。它的特性丰富、选项较多，可以通过以下命令来查看其支持的选项：

```
# qemu-img create -f qcow2 -o ?
Supported options:
size              Virtual disk size
compat            Compatibility level (0.10 or 1.1)
backing_file      File name of a base image
backing_fmt       Image format of the base image
encryption        Encrypt the image with format 'aes'. (Deprecated in favor of encrypt.format=aes)
encrypt.format    Encrypt the image, format choices: 'aes', 'luks'
encrypt.key-secret ID of secret providing qcow AES key or LUKS passphrase
encrypt.cipher-alg Name of encryption cipher algorithm
encrypt.cipher-mode Name of encryption cipher mode
encrypt.ivgen-alg Name of IV generator algorithm
encrypt.ivgen-hash-alg Name of IV generator hash algorithm
encrypt.hash-alg  Name of encryption hash algorithm
encrypt.iter-time Time to spend in PBKDF in milliseconds
cluster_size      qcow2 cluster size
preallocation     Preallocation mode (allowed values: off, metadata, falloc, full)
lazy_refcounts    Postpone refcount updates
refcount_bits     Width of a reference count entry in bits
```

(1) compat：兼容性版本号。默认为 1.1，它比传统格式(版本 0.10)有更高的读取效率。
(2) backing_file：基础映像的文件名。
(3) backing_fmt：基础映像的格式。
(4) encryption：使用 128 位 AES-CBC 加密。
(5) cluster_size：簇大小(必须在 512KB～2MB)。较小的簇可以改善映像文件的大小，而较大的簇可以提供更好的性能。
(6) preallocation：预分配模式，有 4 个可选值：
- off：不使用预分配策略，这是默认选项。
- metadata：为元数据分配空间，映像文件仍然属于稀疏类型。类似于 VMware ESXi 的 Thin Provision 类型的虚拟磁盘。
- full：分配所有的空间并置为零，映像文件属于非稀疏类型。类似于 VMware ESXi 的"厚置备置零"类型的虚拟磁盘。
- falloc：使用 posix_fallocate() 函数分配文件的块并将它们的状态标识为未初始化，创建映像文件的速度要比 full 模式快很多。类似于 VMware ESXi 的"厚置备延迟置零"类型的虚拟磁盘。

(7) lazy_refcounts：仅当指定 compat = 1.1 时才能启用此选项。如果将此选项设置为 on，则将推迟引用计数更新，这可以减少元数据 I/O 并提高性能。这对于 cache = writethrough 模式尤其有用。其缺点是在宿主机崩溃后，必须重新构建引用计数表，即在下一次启动时，会自动执行 qemu-img check -r all 进行修复，这可能需要一些时间。

(8) nocow：仅对 btrfs 文件系统有效。如果将此选项设置为 on，则它将关闭文件的 COW(Copy On Write)特性。

6.3.6 基础映像与派生映像

在生产环境中，很多虚拟机会使用相同版本的操作系统及应用程序，特别是桌面虚拟化系统。可以使用基础映像文件与派生映像文件来减少存储空间的消耗。

先创建一台虚拟机，安装操作系统及应用程序，升级补丁和优化配置之后关机，将这台虚拟机的映像文件当作基础映像文件(有时被称为"黄金映像")，然后基于它创建派生映像文件，派生映像文件仅保存与基础映像文件之间的差异，如图 6-10 所示。

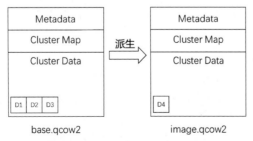

图 6-10 基础映像与派生映像的简化示意图

仅需要将派生映像文件分配给虚拟机，虚拟机会检测到是派生映像文件和基础映像文件"拼接"的结果（有时被称为"视图"），这与 VMware 产品中的链接克隆类似。

可以基于同一个基础映像文件派生出多个映像文件，在 qemu-img 术语中，基础镜像文件被称为 backing_file。

提示：基础映像（Base Image）、父映像（Parent Image）、后备文件（Backing File）含义相同。

基础映像文件既可以是 qcow2 格式也可以是 raw 格式，但是派生映像文件必须是 qcow2 格式文件，还可以基于派生映像文件再派生出新的映像文件，从而形成链状结构，命令如下：

```
1 # qemu-img create -f qcow2 base.qcow2 1G
  Formatting 'base.qcow2', fmt=qcow2 size=1073741824 cluster_size=65536 lazy_refcounts=off refcount_bits=16

2 # qemu-img create -f qcow2 -o backing_file=base.qcow2 image.qcow2
  Formatting 'image.qcow2', fmt=qcow2 size=1073741824 backing_file=base.qcow2 cluster_size=65536 lazy_refcounts=off refcount_bits=16

3 # qemu-img info image.qcow2
  image: image.qcow2
  file format: qcow2
  virtual size: 1.0G (1073741824 Bytes)
  disk size: 196K
  cluster_size: 65536
  backing file: base.qcow2
  Format specific information:
      compat: 1.1
      lazy refcounts: false
      refcount bits: 16
      corrupt: false
```

第 1 行命令创建了映像文件，准备把它当作基础映像。

第 2 行命令创建了一个名为 image.qcow2 的派生映像文件，通过 backing_file=base.qcow2 来指定基础映像文件。

从第 3 行命令的输出可以看出：image.qcow2 的基础映像文件是 base.qcow2。

在初始阶段，image.qcow2 是空的。提供给虚拟机的数据都来自基础映像文件 base.qcow2。当向其中写入新的数据的时候，才会把这些差异的数据保存到 image.qcow2 中，命令如下：

```
4 # qemu-img info --backing-chain image.qcow2
  image: image.qcow2
  file format: qcow2
  virtual size: 1.0G (1073741824 Bytes)
  disk size: 196K
  cluster_size: 65536
  backing file: base.qcow2
  Format specific information:
      compat: 1.1
      lazy refcounts: false
      refcount bits: 16
      corrupt: false

  image: base.qcow2
  file format: qcow2
  virtual size: 1.0G (1073741824 Bytes)
  disk size: 196K
  cluster_size: 65536
  Format specific information:
      compat: 1.1
      lazy refcounts: false
      refcount bits: 16
      corrupt: false
```

第 4 行命令使用了 --backing-chain 选项,这会递归枚举链中每个映像文件的信息。

采用派生映像的优点有以下几点:

(1) 可以快速生成虚拟机镜像。

(2) 可以节省磁盘空间。

(3) 便于维护升级。

采用派生映像的缺点有以下几点:

(1) 性能会有所下降。

(2) 必须保证基础映像的安全,这增加了管理成本。

提示:建议将基础映像文件放置到比较快的存储介质上,例如 SSD 硬盘。

6.3.7 修改映像文件的选项

映像文件创建之后,还可以根据需要使用 amend 子命令修改其选项,其语法格式如下:

```
# qemu-img amend [-p] [-f fmt] [-t cache] -o options filename
```

注意:仅 qcow2 文件格式支持此操作。

6.3.8 转换映像文件格式

可以使用 convert 子命令转换映像格式,其语法格式如下:

```
# qemu-img convert [-c] [-p] [-f fmt] [-t cache] [-O output_fmt] [-o options] [-S sparse_size] filename output_filename
```

既可以在不同映像格式家族之间进行转换(例如:qcow2 转 raw),也可以在不同布局风格之间转换(例如:精简配置格式转换为厚格式,派生映像转换为普通映像)。

执行的命令如下:

```
1 # file test.qcow2
  test.qcow2: QEMU QCOW Image (v3), has backing file (path base.qcow2), 1073741824 Bytes

2 # qemu-img info test.qcow2
  image: test.qcow2
  file format: qcow2
  virtual size: 1.0G (1073741824 Bytes)
  disk size: 196K
  cluster_size: 65536
  backing file: base.qcow2
  Format specific information:
      compat: 1.1
      lazy refcounts: false
      refcount bits: 16
      corrupt: false
```

从第 1 行和第 2 行命令的输出可以看出:test.qcow2 是一个 qcow2 格式的派生映像文件。

接下来执行的命令如下:

```
3 # qemu-img convert -p -O raw test.qcow2 test.raw
      (100.00/100%)

4 # file test.raw
  test.raw: data

5 # qemu-img info test.raw
  image: test.raw
  file format: raw
  virtual size: 1.0G (1073741824 Bytes)
  disk size: 4.0K
```

第 3 行命令对文件 test.qcow2 进行转换。其中-p 参数将显示转换的进度,-O raw 参数用于指定目标文件格式为 raw,新文件名为 test.raw。

第 4 行和第 5 行命令显示了新文件 test.raw 的信息。

6.3.9 比较映像文件

使用 compare 子命令可以对两个映像文件的内容进行对比,其语法格式如下:

```
# qemu-img compare [-f fmt] [-F fmt] [-p] [-s] [-q] imgname1 imgname2
```

命令的退出代码及含义如下。

(1) 0:映像文件相同。

(2) 1:映像文件不同。

(3) 2:打开其中一个映像文件时出错。

(4) 3:检查扇区分配时出错。

(5) 4:读取数据时出错。

执行的命令如下:

```
1 # file test.*
  test.qcow2: QEMU QCOW Image (v3), has backing file (path base.qcow2), 1073741824 Bytes
  test.raw: data

2 # qemu-img compare test.qcow2 test.raw
  Images are identical.
```

第 1 行命令的输出显示出两个文件的格式是不同的。第 2 行命令显示这两个文件中所包含的数据是相同的。

6.3.10 更改基础映像文件

如果更改了基础映像文件的名称或位置,则必须使用 rebase 子命令来修改派生文件。其语法格式如下:

```
# qemu-img rebase [-f fmt] [-t cache] [-p] [-u] -b backing_file [-F backing_fmt] filename
```

使用-b 选项来指定新的基础映像文件。

执行的命令如下:

```
1 # qemu-img info image.qcow2 --backing-chain
  image: image.qcow2
  file format: qcow2
  virtual size: 1.0G (1073741824 Bytes)
```

```
    disk size: 196K
    cluster_size: 65536
    backing file: base.qcow2
    Format specific information:
        compat: 1.1
        lazy refcounts: false
        refcount bits: 16
        corrupt: false

    image: base.qcow2
    file format: qcow2
    virtual size: 1.0G (1073741824 Bytes)
    disk size: 196K
    cluster_size: 65536
    Format specific information:
        compat: 1.1
        lazy refcounts: false
        refcount bits: 16
        corrupt: false
```

从第 1 行命令的输出可以看出：image.qcow2 的基础映像文件是当前目录下的 base.qcow2。

现在需要将基础映像文件移动到另外一个目录中(/vm)，建议采用如下操作：

```
2 # cp base.qcow2 /vm/newbase.qcow2

3 # qemu-img rebase -b /vm/newbase.qcow2 image.qcow2

4 # qemu-img info image.qcow2 --backing-chain
    image: image.qcow2
    file format: qcow2
    virtual size: 1.0G (1073741824 Bytes)
    disk size: 196K
    cluster_size: 65536
    backing file: /vm/newbase.qcow2
    Format specific information:
        compat: 1.1
        lazy refcounts: false
        refcount bits: 16
        corrupt: false

    image: /vm/newbase.qcow2
    file format: qcow2
    virtual size: 1.0G (1073741824 Bytes)
    disk size: 196K
```

```
    cluster_size: 65536
    Format specific information:
        compat: 1.1
        lazy refcounts: false
        refcount bits: 16
        corrupt: false

5 # rm base.qcow2
```

第 2 行命令将基础映像文件复制到新的目录中,这比直接移动文件更稳妥一些。
第 3 行命令将更改 image.qcow2 的基础映像。
第 4 行命令的输出显示:新的基础映像为/vm/newbase.qcow2。
第 5 行命令删除了原有的基础映像。
rebase 子命令有两种不同的工作模式:safe 和 unsafe。
默认为 safe 模式,也是推荐的模式。在进行更改时,会进行一些检查、对比操作,所以要求原有的基础映像文件必须保留。
还可以使用-u 选项来指定为 unsafe 模式。此模式对于重命名或移动备份文件很有用。可以在无法访问旧文件的情况下使用它。例如:如果第 2 行命令不是将文件复制到新的目录下,而是直接移动到新目录中,则第 3 行命令就需要增加-u 选项了。

6.3.11 提交对映像文件的更改

使用 commit 子命令将保存在派生映像文件中的数据全部提交给基础映像。其语法格式如下:

```
# qemu-img commit [-f fmt] [-t cache] imgname
```

实验可参见 6.3.12 节。

警告:如果基于基础映像有多个派生映像,则需要慎用 commit。

6.3.12 显示映像文件布局

使用 map 子命令可以显示映像文件及其基础映像文件(如果有)的元数据,并且会显示映像文件中每个扇区的分配状态。其语法格式如下:

```
# qemu-img map [-f fmt] [--output=fmt] imgname
```

有两种输出格式:human 格式和 json 格式,命令如下:

```
1 #qemu-img info /vm/imag1.qcow2 --backing-chain
  image: /vm/imag1.qcow2
  file format: qcow2
  virtual size: 500M (524288000 Bytes)
  disk size: 25M
  cluster_size: 65536
  backing file: /vm/base.qcow2
  Format specific information:
      compat: 1.1
      lazy refcounts: false
      refcount bits: 16
      corrupt: false

  image: /vm/base.qcow2
  file format: qcow2
  virtual size: 500M (524288000 Bytes)
  disk size: 25M
  cluster_size: 65536
  Format specific information:
      compat: 1.1
      lazy refcounts: false
      refcount bits: 16
      corrupt: false
```

第 1 行命令的输出显示：imag1.qcow2 是 base.qcow2 的派生映像文件。接下来执行的命令如下：

```
2 #qemu-img map /vm/imag1.qcow2
  Offset      Length      Mapped to    File
  0           0x10000     0x50000      /vm/imag1.qcow2
  0x10000     0x30000     0x60000      /vm/base.qcow2
  0x40000     0x20000     0x70000      /vm/imag1.qcow2
  0x60000     0x20000     0xb0000      /vm/base.qcow2
  0x80000     0x30000     0x110000     /vm/base.qcow2
  …

3 #qemu-img map /vm/imag1.qcow2 | grep imag1.qcow2 | wc -l
  6

4 #qemu-img map /vm/imag1.qcow2 | grep base.qcow2 | wc -l
  73
```

第 2 行命令的输出显示了映像文件的布局。默认格式为 human，它会显示文件中已经使用的、非零的部分。每行包括 4 个字段。以第一行为例，它表示从偏移量 0 开始的、长度是 0x10000 字节的数据，对应的是在文件/vm/imag1.qcow2 中从 0x50000 开始的偏移量。

通过第 3 行、第 4 行的简单统计可以看到,6 块 imag1.qcow2 和 73 块 base.qcow2 中连续的扇区共同组成了虚拟磁盘中的数据。

接下来执行的命令如下:

```
5 #qemu-img commit /vm/imag1.qcow2

6 #qemu-img map /vm/imag1.qcow2
  Offset          Length          Mapped to        File
  0               0x80000         0x50000          /vm/base.qcow2
  0x80000         0x30000         0x110000         /vm/base.qcow2
  0xb0000         0x40000         0xd0000          /vm/base.qcow2
  0xf0000         0x2b0000        0x140000         /vm/base.qcow2
  …

7 #qemu-img map /vm/imag1.qcow2 | grep imag1.qcow2 | wc -l
  0

8 #qemu-img map /vm/imag1.qcow2 | grep base.qcow2 | wc -l
  82
```

第 5 行命令将保存在派生映像文件中的数据提交到基础映像文件中。

第 6 行命令的输出显示:所有非零数据都被保存到基础映像文件/vm/base.qcow2 中。这也可以通过第 7 行、第 8 行的简单统计结果得到验证。

6.3.13 快照管理

可以使用 snapshot 子命令管理 qcow2 格式映像的快照。其语法格式如下:

```
#qemu-img snapshot [ -l | -a snapshot | -c snapshot | -d snapshot ] filename
```

常用的选项及参数有以下几个。

(1) -a:应用快照(将映像文件还原到指定的快照状态)。
(2) -c:创建快照。
(3) -d:删除快照。
(4) -l:列出映像文件的所有快照。
(5) snapshot:此参数指定要创建、应用或删除的快照名称(标记)。

执行的命令如下:

```
1 #qemu-img info /vm/test.qcow2
  image: /vm/test.qcow2
  file format: qcow2
  virtual size: 9.0G (9663676416 Bytes)
```

```
        disk size: 937M
        cluster_size: 65536
        Format specific information:
            compat: 1.1
            lazy refcounts: false
            refcount bits: 16
            corrupt: false
```

第1行命令的输出显示了映像文件/vm/test.qcow2没有快照。

接下来执行的命令如下:

```
2 # qemu-img snapshot -c snapshot1 /vm/test.qcow2

3 # qemu-img snapshot -l /vm/test.qcow2
  Snapshot list:
  ID        TAG             VM SIZE         DATE            VM CLOCK
  1         snapshot1       0 2020-12-26    22:12:42        00:00:00.000
4 # qemu-img snapshot -c snapshot2 /vm/test.qcow2

5 # qemu-img snapshot -l /vm/test.qcow2
  Snapshot list:
  ID        TAG             VM SIZE         DATE            VM CLOCK
  1         snapshot1       0 2020-12-26    22:12:42        00:00:00.000
  2         snapshot2       0 2020-12-26    22:13:00        00:00:00.000
```

第2行命令通过-c选项指定了新快照的名称为snapshot1。

第3行命令输出了此映像文件的快照列表。一行表示一个快照,包括快照的ID、名称、大小、创建的日期与时间。

还可以通过qemu-img info命令获得映像文件的快照列表,命令如下:

```
6 # qemu-img info /vm/test.qcow2
  image: /vm/test.qcow2
  file format: qcow2
  virtual size: 9.0G (9663676416 Bytes)
  disk size: 938M
  cluster_size: 65536
  Snapshot list:
  ID        TAG             VM SIZE         DATE            VM CLOCK
  1         snapshot1       0 2020-12-26    22:12:42        00:00:00.000
  2         snapshot2       0 2020-12-26    22:13:00        00:00:00.000
  Format specific information:
      compat: 1.1
      lazy refcounts: false
      refcount bits: 16
      corrupt: false
```

接下来执行的命令如下:

```
7 # qemu-img snapshot -a 1 /vm/test.qcow2
```

第 7 行命令将映像文件恢复到快照 ID 是 1 的状态。
接下来执行的命令如下:

```
8 # qemu-img snapshot -d 2 /vm/test.qcow2
```

```
9 # qemu-img snapshot -d snapshot1 /vm/test.qcow2
```

第 8 行命令通过快照 ID 删除指定快照,第 9 行命令则通过快照名称进行指定。
接下来执行的命令如下:

```
10 # qemu-img info /vm/test.qcow2
   image: /vm/test.qcow2
   file format: qcow2
   virtual size: 9.0G (9663676416 Bytes)
   disk size: 937M
   cluster_size: 65536
   Format specific information:
       compat: 1.1
       lazy refcounts: false
       refcount bits: 16
       corrupt: false
```

第 10 行命令显示此映像文件的当前状态,现在已经没有快照了。

6.4 存储池

存储池(Storage Pool)是由 libvirt 管理的为虚拟机预留的宿主机上的存储容量。可以在存储池的空间中创建存储卷(Storage Volume),然后将存储卷当作块设备分配给虚拟机。

由于是通过存储池和存储卷的名称将存储分配给虚拟机的,虚拟机并不需要知道存储的底层物理路径,所以这就提高了系统管理的灵活性。

Cockpit、virt-manager、virsh 等工具都是通过 libvirt API 来管理虚拟存储的。其中 virsh 功能最强大,有些操作仅能通过它实现,例如:只有 virsh 命令可以创建临时存储池。

virsh 有多个以 pool 开头的并与存储池有关的子命令,如表 6-2 所示。

表 6-2　virsh 中存储池管理子命令

virsh 子命令	功能说明
pool-autostart	将存储池设置为自动启动
pool-build	构建一个存储池

virsh 子命令	功能说明
pool-create-as	通过一组命令参数来创建并启动临时存储池(无配置文件)
pool-create	通过 XML 文件来创建并启动临时存储池(无配置文件)
pool-define-as	通过一组命令行参数来定义存储池(非活动的)
pool-define	通过 XML 文件来定义存储池(非活动的),或修改现有的存储池
pool-delete	删除一个存储池
pool-destroy	停止一个存储池
pool-dumpxml	以 XML 格式显示存储池的信息
pool-edit	编辑存储池的 XML 配置
pool-info	显示存储池的信息
pool-list	显示存储池的列表
pool-name	将存储池的 UUID 转换为名称
pool-refresh	刷新存储池
pool-start	启动一个已定义的非活动的存储池
pool-undefine	取消一个不活动的存储池定义
pool-uuid	将存储池的名称转换为 UUID
pool-event	显示存储池的事件

6.4.1 查看当前存储池

执行的命令如下:

```
1 #virsh pool-list --all --details
  Name     State    Autostart  Persistent  Capacity   Allocation  Available
  -------------------------------------------------------------------------
  default  running  yes        yes         16.99 GiB  10.31 GiB   6.68 GiB
  iso      running  yes        yes         16.99 GiB  10.31 GiB   6.68 GiB
  vm       running  yes        yes         79.96 GiB  14.70 GiB   65.26 GiB
```

第 1 行命令列出了此宿主机上所有存储池的详细信息,包括活动及非活动的存储池,以及每个存储池的状态、是否自动启动、是否是永久性存储池、总容量、已经分配容量及可用容量等信息。

接下来执行的命令如下:

```
2 #virsh pool-info default
  Name:        default
  UUID:        03c7c0ae-8ad7-41ef-9820-04e9f0a0e2e7
  State:       running
  Persistent:  yes
  Autostart:   yes
```

```
    Capacity:        16.99 GiB
    Allocation:      10.31 GiB
    Available:       6.68 GiB

3 # virsh pool-dumpxml default
  <pool type='dir'>
    <name>default</name>
    <uuid>03c7c0ae-8ad7-41ef-9820-04e9f0a0e2e7</uuid>
    <capacity unit='Bytes'>18238930944</capacity>
    <allocation unit='Bytes'>11067076608</allocation>
    <available unit='Bytes'>7171854336</available>
    <source>
    </source>
    <target>
      <path>/var/lib/libvirt/images</path>
      <permissions>
        <mode>0711</mode>
        <owner>0</owner>
        <group>0</group>
        <label>system_u:object_r:virt_image_t:s0</label>
      </permissions>
    </target>
  </pool>
```

第 2 行、第 3 行输出了存储池 default 的详细信息。其中<pool type＝'dir'>表示这是一个基于目录的存储池。

接下来执行的命令如下：

```
4 # tree /etc/libvirt/storage/
  /etc/libvirt/storage/
  ├── autostart
  │   ├── default.xml -> /etc/libvirt/storage/default.xml
  │   ├── iso.xml -> /etc/libvirt/storage/iso.xml
  │   └── vm.xml -> /etc/libvirt/storage/vm.xml
  ├── default.xml
  ├── iso.xml
  └── vm.xml
```

每个存储池的 XML 定义文件都保存在/etc/libvirt/storage/目录中。不建议直接修改此目录中的文件。第 4 行命令显示了此目录的文件及子目录。如果在 autostart 子目录有同名的符号链接文件，则说明它是自动启动的存储池。

6.4.2 存储池的分类

从不同的视角可以将存储池划分成不同的类。

1．根据存储池的实效性划分

（1）持久性存储池：在宿主机重新引导后仍然有效。

（2）临时存储池：仅在宿主机当前运行时有效，重新引导之后会消失。

提示：只有 virsh 命令可以创建临时存储池。

2．根据底层存储技术来划分

（1）基于目录的存储池。

（2）基于磁盘的存储池。

（3）基于分区的存储池。

（4）基于 GlusterFS 的存储池。

（5）基于 iSCSI 的存储池。

（6）基于 LVM 的存储池。

（7）基于 NFS 的存储池。

（8）基于 vHBA 的存储池（具有 SCSI 设备的）。

（9）基于多路径的存储池。

（10）基于 RBD 的存储池。

（11）基于 Sheepdog 的存储池。

（12）基于 Vstorage 的存储池。

（13）基于 ZFS 的存储池。

提示：RHEL/CentOS 8 不支持基于多路径、RBD、Sheepdog、Vstorage 和 ZFS 的存储池。

3．根据数据存储的位置划分

（1）本地存储池：使用的是直接连接到宿主机上的存储。

（2）网络（共享）存储池：通过标准协议使用网络上共享的存储设备。

6.4.3 创建存储池的通用流程

Cockpit、virt-manager 简单易用，但只能创建常用的存储池，例如：Cockpit 仅支持以下 5 种存储池的创建：

（1）基于目录的存储池。

（2）基于物理磁盘的存储池。

（3）基于 LVM 卷组的存储池。

（4）基于网络文件系统的存储池。

（5）基于 iSCSI 目标的存储池。

virsh 虽然操作复杂一些，但是功能最为强大。通过 virsh 创建存储池主要有 4 步。

1）定义（define）存储池

有两种定义方法：

（1）使用 XML 文件定义存储池。
- 使用 virsh pool-define 命令创建持久性存储池（生成配置文件）。
- 使用 virsh pool-create 命令创建并启动临时存储池（无配置文件）。

（2）使用参数定义存储池。
- 使用 virsh pool-define-as 命令创建持久性存储池（生成配置文件）。
- 使用 virsh pool-create-as 命令创建并启动临时存储池（无配置文件）。

2）构建（build）存储池的目标路径

使用 virsh pool-build 命令来构建目标路径。根据存储池类型的不同，操作结果会有所差异：为基于目录和基于网络文件系统的存储池创建目标路径；为基于磁盘的存储池初始化磁盘；为基于 LVM 卷组的存储池创建卷组。

3）启动（start）存储池

使用 virsh pool-start 命令准备要使用的源设备。根据存储池类型的不同，操作结果会有所差异：如果是基于目录的存储池，则会根据需要创建新目录；如果是基于文件系统的存储池，则会进行 mount 操作；如果是基于 LVM 的存储池，则会创建 LVM 的卷组；如果是基于 NFS 的存储池，则会进行 mount 操作；如果是基于 iSCSI 的存储池，则会发起到 iSCSI Target 的连接……

4）将存储池设置为自动启动（autostart）

这是可选的操作。使用 virsh pool-autostart 命令，可以将存储池设置为随 libvirtd 守护程序的启动而自动启动。

6.4.4 基于目录的存储池

基于目录的存储池是最常见的存储池类型。它的容器就是文件系统的目录，其中的存储卷就是目录中的映像文件。

下面的实验将通过 virsh 来创建一个基于目录的持久性存储池。

首先创建一个 XML 文件，用于指定新存储池的参数，命令如下：

```
1 # vi new.xml
  < pool type = 'dir'>
    < name > vm2 </name >
    < target >
      < path >/vm2 </path >
    </target >
  </pool >
```

参数的含义如表 6-3 所示。

表 6-3　基于目录的存储池的参数

描述	XML 格式
存储池的类型	<pool type='dir'>
存储池的名称	<name>name</name>
存储池的目标路径	<target> 　<path>target_path</path> </target>

接下来执行的命令如下：

```
2 #virsh pool-define new.xml
  Pool vm2 defined from new.xml

3 #virsh pool-list --all
  Name      State      Autostart
  ---------------------------------------------
  default   active     yes
  iso       active     yes
  vm        active     yes
  vm2       inactive   no
```

第 2 行命令仅仅创建了新存储池的定义。从第 3 行命令的输出中可以看出：新存储池并没有启动，而是处于非活动状态。

也可以通过向 pool-define-as 子命令传递参数来创建存储池的定义。这种方式比较适合在命令行或脚本中使用，命令如下：

```
#virsh pool-define-as --name vm2 --type dir --target "/vm2"
```

接下来执行的命令如下：

```
4 #ls -t /etc/libvirt/storage/
  vm2.xml autostart vm.xml iso.xml default.xml

5 #cat /etc/libvirt/storage/vm2.xml
  <!--
  WARNING: THIS IS AN AUTO-GENERATED FILE. CHANGES TO IT ARE LIKELY TO BE
  OVERWRITTEN AND LOST. Changes to this xml configuration should be made using:
    virsh pool-edit vm2
  or other application using the libvirt API.
  -->

  <pool type='dir'>
    <name>vm2</name>
```

```
      <uuid>eefab204-bee6-462e-8b78-3baa91e188c6</uuid>
      <capacity unit='Bytes'>0</capacity>
      <allocation unit='Bytes'>0</allocation>
      <available unit='Bytes'>0</available>
      <source>
      </source>
      <target>
        <path>/vm2</path>
      </target>
    </pool>
```

第 4 行命令的输出显示：在/etc/libvirt/storage/目录中新增了一个与存储池同名的 XML 文件，第 5 行命令显示了这个 XML 文件的内容。

接下来执行的命令如下：

```
6 # ls -ld /vm2
  ls: cannot access '/vm2': No such file or directory

7 # virsh pool-build vm2
  Pool vm2 built

8 # ls -ld -Z /vm2
  drwx--x--x. 2 root root system_u:object_r:root_t:s0 6 Dec 25 16:53 /vm2
```

此时 libvirt 并不会为这个基于目录的存储池创建目标目录结构。当通过第 7 行 pool-build 子命令构建存储池时，如果目标目录结构(/vm2)不存在，则 libvirt 才会创建目标目录。

使用第 8 行命令查看新目录的信息，包括 SELinux 的上下文。

提示：RHEL/CentOS 8 中/var/lib/libvirt/images 目录的 SELinux 下上文是：system_u:object_r:virt_image_t:s0。

接下来执行的命令如下：

```
9 # virsh pool-start vm2
  Pool vm2 started

10 # virsh pool-list --all
   Name       State      Autostart
   -----------------------------------------
   default    active     yes
   iso        active     yes
   vm         active     yes
   vm2        active     no
```

第 9 行命令用于启动存储池 vm2。只有当存储池启动之后,虚拟机才能访问其中的存储卷。

第 10 行命令的输出显示:存储池 vm2 的状态为 active。

接下来执行的命令如下:

```
11 # virsh pool-autostart vm2
   Pool vm2 marked as autostarted

12 # virsh pool-list --all
   Name          State       Autostart
   -------------------------------------------
   default       active      yes
   iso           active      yes
   vm            active      yes
   vm2           active      yes
```

第 11 行命令将存储池 vm2 设置为自动启动。第 12 行命令的输出显示:存储池 vm2 的 Autostart 属性是 yes。

接下来执行的命令如下:

```
13 # virsh pool-dumpxml vm2
   <pool type='dir'>
     <name>vm2</name>
     <uuid>eefab204-bee6-462e-8b78-3baa91e188c6</uuid>
     <capacity unit='Bytes'>18238930944</capacity>
     <allocation unit='Bytes'>11066273792</allocation>
     <available unit='Bytes'>7172657152</available>
     <source>
     </source>
     <target>
       <path>/vm2</path>
       <permissions>
         <mode>0711</mode>
         <owner>0</owner>
         <group>0</group>
         <label>system_u:object_r:root_t:s0</label>
       </permissions>
     </target>
   </pool>
```

第 13 行命令显示了存储池 vm2 的详细信息,包括存储池类型、名称、UUID、总容量、已分配容量、可用容量、目标路径、权限、SELinux 上下文等。

提示:基于目录的存储池的总容量、已分配容量和可用容量是目录所有分区或卷的总容量、已分配容量和可用容量。

当没有虚拟机使用存储池中的存储卷时，就可以将其删除，命令如下：

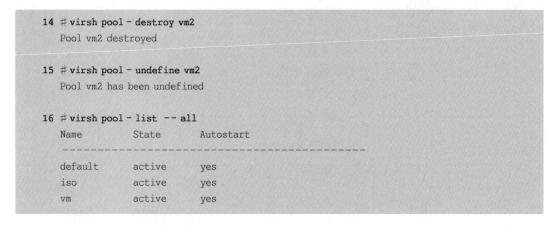
```
14 # virsh pool-destroy vm2
   Pool vm2 destroyed

15 # virsh pool-undefine vm2
   Pool vm2 has been undefined

16 # virsh pool-list --all
 Name              State      Autostart
-------------------------------------------
 default           active     yes
 iso               active     yes
 vm                active     yes
```

第 14 行命令用于停止存储池 vm2，它将处于 inactive 状态。

第 15 行命令取消了这个不活动的存储池的定义，同时也会删除 /etc/libvirt/storage/ 目录中对应的 XML 文件。

提示：删除存储池并不会自动删除目标目录及其中的映像文件。

在 Cockpit 中创建基于目录的存储池的操作很简单。单击位于 Virtual Machines 顶部的 Storage Pools 链接，会出现 Storage Pools 窗口。其中显示了已配置的存储池的列表，单击 Create Storage Pool 链接，在新的对话框中输入新存储池的信息，如图 6-11 所示。

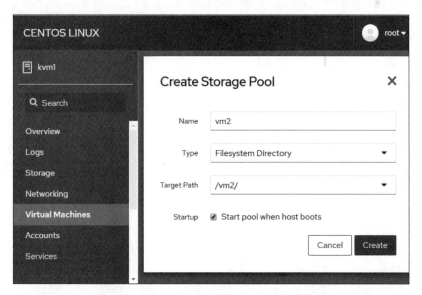

图 6-11　在 Cockpit 中创建基于目录的存储池

(1) Name：存储池的名称。

(2) Type：存储池的类型，可以是文件系统目录、网络文件系统、iSCSI 目标、物理磁盘驱动器或 LVM 卷组。本次选择 Filesystem Directory。

(3) Target Path：宿主机文件系统上的存储池路径。

(4) Startup：宿主机启动时是否启动存储池。

单击 Create 按钮将创建存储池，新的存储池会出现在存储池列表中。

提示：在 Cockpit 中创建基于目录的存储池时，并不会创建新的目标目录，应当事先在宿主机上创建好目录。

在 virt-manager 中创建基于目录的存储池的操作与 Cockpit 类似。在 virt-manager 中，选择要配置的宿主机连接。打开 Edit 菜单，然后选择 Connection Details。单击 Connection Details 窗口中的 Storage 选项卡。单击窗口底部的 按钮，就会出现 Add a New Storage Pool 窗口，如图 6-12(a)所示。

输入新存储池的名称。在 Type 下拉列表中选择存储池类型，本示例使用 dir：Filesystem Directory。

单击 XML 选项卡，可以显示新存储池的 XML 定义，如图 6-12(b)所示。

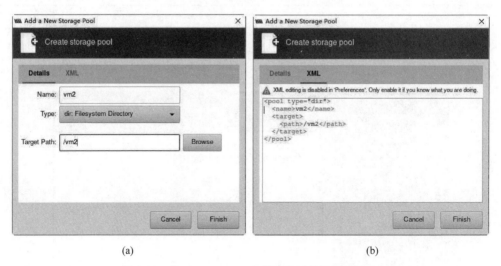

图 6-12　在 virt-manager 中创建基于目录的存储池

单击 Finish 按钮创建存储池，新的存储池就会出现在存储池列表中。

提示：virt-manager 会自动创建新的目标目录。与 Cockpit 相同，只能创建持久性的存储池。如果希望创建临时存储池，就需要使用 virsh 的 pool-create 或 pool-create-as 子命令。

6.4.5 基于物理磁盘的存储池

基于物理磁盘的存储池的容器是整个磁盘,其中的存储卷就是分区。由于直接将分区分配给虚拟机,所以没有文件系统的开销,比较适用于对性能要求高的场景,但是管理的灵活性就差一些,例如:无快照功能、存储卷数量有限(GPT 磁盘最多可以分为 128 个分区)。

下面先通过 virsh 命令创建一个基于物理磁盘的持久性的存储池。

在实验环境中给宿主机添加一个 107GB 的新磁盘,命令如下:

```
1 # parted /dev/sdc print
  Error: /dev/sdc: unrecognised disk label
  Model: VMware, VMware Virtual S (scsi)
  Disk /dev/sdc: 107GB
  Sector size (logical/physical): 512B/512B
  Partition Table: unknown
  Disk Flags:
```

新磁盘的设备名称是/dev/sdc。从第 1 行命令的输出中可以看出此磁盘没有分区表。接下来执行的命令如下:

```
2 # vi new.xml
  <pool type='disk'>
    <name>vm2</name>
    <source>
      <device path='/dev/sdc'/>
      <format type='gpt'/>
    </source>
    <target>
      <path>/dev</path>
    </target>
  </pool>
```

第 2 行命令创建了一个 XML 文件,用于指定新存储池的参数,参数说明如表 6-4 所示。

表 6-4 基于物理磁盘存储池的参数

描述	XML 格式
存储池的类型	<pool type='disk'>
存储池的名称	<name>name</name>
源存储设备的路径及格式	<source> <path>source_path</path> <format type='source_type'/> </source>
存储池的目标路径,对于基于物理磁盘的存储池通常是/dev	<target> <path>target_path</path> </target>

在 XML 定义中,通过<format type='gpt'/>来指定分区表的类型。如果是全新的磁盘,在后续执行 pool-build 子命令时,则会根据这个值来创建分区表,推荐使用 GPT 格式的分区表。如果是已经分过区的磁盘,则需要设置与分区表格式相同的值,例如 mbr 或 gpt。

接下来执行的命令如下:

```
3 # virsh pool-define new.xml
  Pool vm2 defined from new.xml

4 # virsh pool-list --all --details
  Name          State      Autostart   Persistent   Capacity    Allocation   Available
  ---------------------------------------------------------------------------------------
  default       running    yes         yes          49.98 GiB   47.63 GiB    2.34 GiB
  iso           running    yes         yes          49.98 GiB   47.63 GiB    2.34 GiB
  virtio-win    running    yes         yes          49.98 GiB   47.63 GiB    2.34 GiB
  vm            running    yes         yes          99.95 GiB   42.62 GiB    57.32 GiB
  vm2           inactive   no          yes
```

第 3 行命令会创建存储池的定义。

从第 4 行命令的输出可以看出:新的存储池处于非活动状态。由于还未初始化,所以未显示容量信息。

接下来执行的命令如下:

```
5 # virsh pool-build vm2
  Pool vm2 built

6 # parted /dev/sdc print
  Model: VMware, VMware Virtual S (scsi)
  Disk /dev/sdc: 107GB
  Sector size (logical/physical): 512B/512B
  Partition Table: gpt
  Disk Flags:
```

第 5 行命令将构建存储池。libvirt 在构建时会检查磁盘的分区情况,如果是全新的磁盘,则会创建指定类型的分区表。从第 6 行的输出可以看出:/dev/sdc 的分区表格式是 GPT。

接下来执行的命令如下:

```
7 # virsh pool-start vm2

8 # virsh pool-autostart vm2

9 # virsh pool-list --all --details
```

Name	State	Autostart	Persistent	Capacity	Allocation	Available
default	running	yes	yes	49.98 GiB	47.63 GiB	2.34 GiB
iso	running	yes	yes	49.98 GiB	47.63 GiB	2.34 GiB
virtio-win	running	yes	yes	49.98 GiB	47.63 GiB	2.34 GiB
vm	running	yes	yes	99.95 GiB	42.62 GiB	57.32 GiB
vm2	running	yes	yes	100.00 GiB	0.00 B	100.00 GiB

第 7 行命令用于启动存储池，第 8 行命令又将其设置为自动启动。

第 9 行命令显示了当前存储池的列表。

接下来执行的命令如下：

```
10 # virsh pool-info vm2
   Name:        vm2
   UUID:        9d682c3e-6727-4828-937c-31503c8ad591
   State:       running
   Persistent:  yes
   Autostart:   yes
   Capacity:    100.00 GiB
   Allocation:  0.00 B
   Available:   100.00 GiB

11 # virsh pool-dumpxml vm2
   <pool type='disk'>
     <name>vm2</name>
     <uuid>9d682c3e-6727-4828-937c-31503c8ad591</uuid>
     <capacity unit='Bytes'>107374165504</capacity>
     <allocation unit='Bytes'>0</allocation>
     <available unit='Bytes'>107374148096</available>
     <source>
       <device path='/dev/sdc'>
         <freeExtent start='17408' end='107374165504'/>
       </device>
       <format type='gpt'/>
     </source>
     <target>
       <path>/dev</path>
     </target>
   </pool>
```

第 10 行、第 11 行命令以不同的格式输出了存储池 vm2 的详细信息。

当没有虚拟机使用存储池中的存储卷时，就可以将其删除。方法是先使用 pool-destroy 子命令停止该存储池，然后使用 pool-undefine 子命令取消存储池的定义。删除基于物理磁盘的存储池，并不会自动删除磁盘上的分区结构及数据。

当前版本 Cockpit(cockpit-machines-211.3-1)在创建基于物理磁盘的存储池时，并不会初始化磁盘(相当于没有 pool-build 操作)。如果要使用全新的磁盘，就需要先使用分区工具创建好分区表，推荐使用 GPT 格式分区表，然后在 Format 下拉列表框中选择正确的分区表格式，如图 6-13 所示。

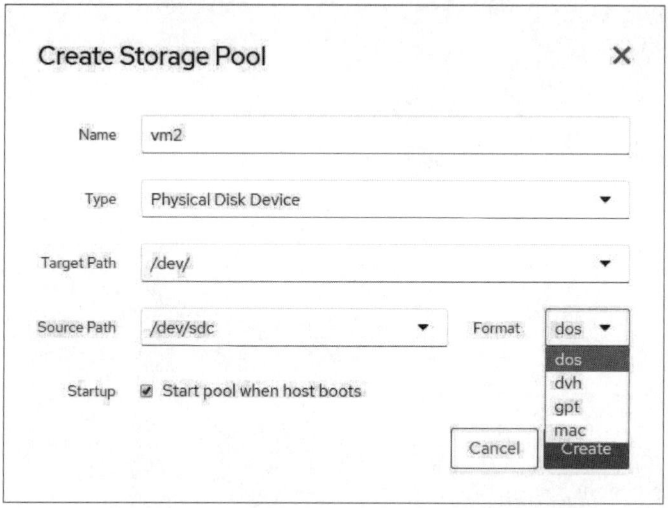

图 6-13　在 Cockpit 中创建基于物理磁盘的存储池

在 virt-manager 中创建基于物理磁盘的存储池时，可以选中 Build Pool 以创建分区表，如图 6-14 所示。

图 6-14　在 virt-manager 中创建基于物理磁盘的存储池(1)

单击 XML 选项卡，可以显示新存储池的 XML 定义，如图 6-15 所示。

单击 Finish 按钮创建存储池。新的存储池会出现在存储池列表中。

图 6-15　在 virt-manager 中创建基于物理磁盘的存储池（2）

6.4.6　基于 LVM 卷组的存储池

逻辑卷管理是 Linux 上最灵活、使用最广泛的存储管理技术。基于 LVM 卷组的存储池的容器是卷组，其中的存储卷就是 LVM 逻辑卷。

libvirt 既可以使用现有的 LVM 卷组构建存储池，也可以通过创建全新的 LVM 卷组实现存储池。

下面将通过 virsh 来创建使用新卷组的存储池，它是由两块物理磁盘组成的 LVM 卷组，命令如下：

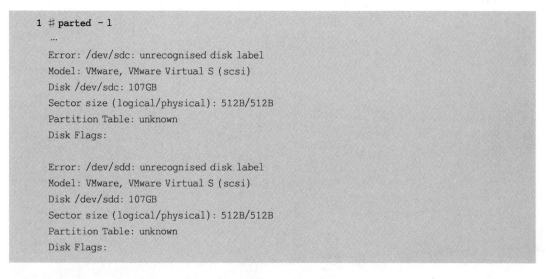

/dev/sdc 和/dev/sdd 是两个未使用的 107GB 的磁盘。

接下来执行的命令如下：

```
2 # vi new.xml
  < pool type = "logical">
    < name > vm2 </name >
    < source >
      < device path = '/dev/sdc'/>
      < device path = '/dev/sdd'/>
    </source >
    < target >
      < path >/dev </path >
    </target >
  </pool >
```

第2行命令创建了一个新 XML 文件，它保存了新存储池的参数，参数说明如表 6-5 所示。

表 6-5　基于 LVM 卷组的存储池的参数

描　　述	XML 格式
存储池的类型	< pool type= 'logical'>
存储池的名称	< name > name </name >
源存储设备的路径	< source > < device path= 'device_path' />
LVM 卷组的名称	< name > VGname </name >
LVM 卷组的格式	< format type= 'lvm2' /></source >
存储池的目标路径	< target > < path= target_path /></target >

如果需要创建全新的 LVM 卷组，则可以通过多个< device path = 'device_path' />来指定多个物理卷，它们既可以是整个磁盘，也可以是分区。如果使用现有的 LVM 卷组，则需要通过< name > VGname </name >指定卷组名称，通过< format type = 'lvm2' />来指定 LVM 版本。

接下来执行的命令如下：

```
3 # virsh pool - define new.xml
  Pool vm2 defined from new.xml

4 # virsh pool - list -- all -- details
  Name          State      Autostart   Persistent   Capacity    Allocation   Available
  ---------------------------------------------------------------------------------------
  default       running    yes         yes          49.98 GiB   47.63 GiB    2.34 GiB
  iso           running    yes         yes          49.98 GiB   47.63 GiB    2.34 GiB
  virtio - win  running    yes         yes          49.98 GiB   47.63 GiB    2.34 GiB
  vm            running    yes         yes          99.95 GiB   42.62 GiB    57.32 GiB
  vm2           inactive   no          yes          —           —            —
```

第 3 行命令创建了存储池的定义。

从第 4 行命令的输出可以看出：新的存储池处于非活动状态。由于还未初始化，所以未显示容量信息。

接下来执行的命令如下：

```
5 # virsh pool-build vm2
  Pool vm2 built

6 # pvs
  PV         VG    Fmt   Attr  PSize      PFree
  /dev/sda2  cl    lvm2  a--   <79.00g    0
  /dev/sdb1  vmvg  lvm2  a--   <100.00g   0
  /dev/sdc   vm2   lvm2  a--   <100.00g   <100.00g
  /dev/sdd   vm2   lvm2  a--   <100.00g   <100.00g

7 # vgs
  VG    #PV  #LV  #SN  Attr    VSize      VFree
  cl    1    3    0    wz--n-  <79.00g    0
  vm2   2    0    0    wz--n-  199.99g    199.99g
  vmvg  1    1    0    wz--n-  <100.00g   0
```

第 5 行命令将构建存储池。由于这是基于 LVM 卷组的存储池，所以 libvirt 会创建 LVM 的物理卷并组成卷组。

从第 6 行命令的输出可以看出：新增加了两个物理卷(/dev/sdc 和/dev/sdd)，它们都属于卷组 vm2。

第 7 行命令输出了宿主机上 LVM 卷组的信息。

接下来执行的命令如下：

```
8 # virsh pool-start vm2

9 # virsh pool-autostart vm2

10 # virsh pool-list --all --details
   Name         State    Autostart  Persistent  Capacity    Allocation  Available
   -----------------------------------------------------------------------------
   default      running  yes        yes         49.98 GiB   47.63 GiB   2.34 GiB
   iso          running  yes        yes         49.98 GiB   47.63 GiB   2.34 GiB
   virtio-win   running  yes        yes         49.98 GiB   47.63 GiB   2.34 GiB
   vm           running  yes        yes         99.95 GiB   42.62 GiB   57.32 GiB
   vm2          running  yes        yes         199.99 GiB  0.00 B      199.99 GiB
```

第 8 行命令用于启动存储池，第 9 行命令又将其设置为自动启动。

第 10 行命令显示了当前存储池的列表。新存储池 vm2 的容量就是 LVM 卷组的容量。

删除基于 LVM 卷组的存储池的方法,还是先使用 pool-destroy 子命令停止该存储池,然后使用 pool-undefine 子命令取消存储池的定义。删除基于 LVM 卷组的存储池,并不会自动删除 LVM 的卷组、逻辑卷及数据。

当前版本 Cockpit(cockpit-machines-211.3-1)只能使用现存的 LVM 卷组来创建存储池,如图 6-16 所示。

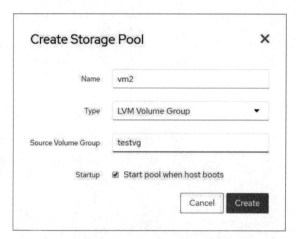

图 6-16 在 Cockpit 中创建基于 LVM 卷组的存储池

在 virt-manager 中的操作也类似,如图 6-17(a)所示。单击 XML 选项卡,可以显示新存储池的 XML 定义,如图 6-17(b)所示。

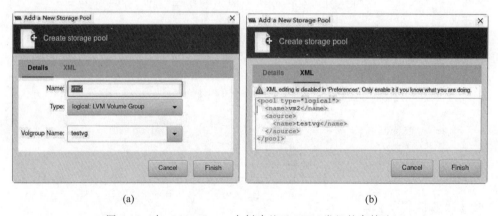

图 6-17 在 virt-manager 中创建基于 LVM 卷组的存储池

6.4.7 基于网络文件系统的存储池

在生产环境中,通常应当将虚拟机的存储保存在集中的外部存储设备上,而不是放置在宿主机的本地存储中。基于网络文件系统的存储池就是一种常见的集中存储解决方案。

常用的网络文件系统有 NFS 和 CIFS,一旦将这些共享的卷挂载到宿主机的目录上,其管理操作与本地文件系统是类似的。下面就以 NFS 来做实验。

在实验中,会部署一台运行 CentOS 8 的主机(主机名为 stor1,IP 地址是 192.168.1.233)。在其中安装 NFS 服务器端组件,它为虚拟化宿主机提供了 NFS 共享,命令如下:

```
1 [root@stor1 ~]# cat /etc/redhat-release
  CentOS Linux release 8.2.2004 (Core)

2 [root@stor1 ~]# yum -y install nfs-utils

3 [root@stor1 ~]# systemctl enable nfs-server

4 [root@stor1 ~]# mkdir /vmdata

5 [root@stor1 ~]# chmod a+w /vmdata

6 [root@stor1 ~]# echo "/vmdata *(rw,no_root_squash,sync)" >> /etc/exports

7 [root@stor1 ~]# systemctl restart nfs-server
```

NFS 服务器端组件包含在 nfs-utils 软件包中。通过第 2 行命令进行安装。将 /vmdata 作为共享的目录,并设置适当的权限。

接下来执行的命令如下:

```
8 [root@stor1 ~]# firewall-cmd --permanent --add-service={nfs,nfs3,rpc-bind,mountd}

9 [root@stor1 ~]# firewall-cmd --reload

10 [root@stor1 ~]# firewall-cmd --list-all
   public (active)
     target: default
     icmp-block-inversion: no
     interfaces: ens32
     sources:
     services: cockpit dhcpv6-client mountd nfs nfs3 rpc-bind ssh
     ports:
     protocols:
     masquerade: no
     forward-ports:
     source-ports:
     icmp-blocks:
     rich rules:
```

```
11 [root@stor1 ~]# showmount -e localhost
   Export list for localhost:
   /vmdata *
```

设置防火墙的规则以允许 NFS 入站的流量。在 NFS 服务器上执行第 11 行命令进行自我检查，要保证 NFS 服务器端配置正确。

提示：建议配置更严格的 NFS 权限及防火墙规则，仅允许来自虚拟化宿主机的访问。

在创建存储池之前，要保证虚拟化宿主机可以正确地访问 NFS 共享目录，下面就做一些简单的测试，命令如下：

```
12 # showmount -e 192.168.1.233
   Export list for 192.168.1.233:
   /vmdata *

13 # mkdir /testfs

14 # mount -t nfs 192.168.1.233:/vmdata /testfs/

15 # echo "test" > /testfs/1.txt

16 # cat /testfs/1.txt
   test

17 # rm /testfs/1.txt

18 # umount /testfs

19 # rmdir /testfs/
```

测试通过之后就可以创建 XML 文件了，它将用于设置新存储池的参数。参数的含义如表 6-6 所示，执行的命令如下：

```
20 # vi new.xml
   <pool type="netfs">
     <name>vm2</name>
     <source>
       <format type="auto"/>
       <host name="192.168.1.233"/>
       <dir path="/vmdata"/>
     </source>
     <target>
       <path>/vm2</path>
     </target>
   </pool>
```

表 6-6　基于网络文件系统的存储池的参数

描　述	XML 格式
存储池的类型	< pool type＝'netfs'>
存储池的名称	< name > name </name >
网络文件服务器的主机名或 IP 地址	< source >< host name＝hostname/>
存储池的格式	可以是以下格式之一：
	< format type＝'nfs' />
	< format type＝'glusterfs' />
	< format type＝'cifs' />
网络文件服务器上的共享目录	< dir path＝source_path /></source >
存储池的目标路径	< target >
	< path > target_path </path >
	</target >

在 XML 定义中，通过< target >< path >/vm2 </path ></target >指定宿主机的目标路径，如果路径目录不存在，则在进行 pool-build 时会自动创建，命令如下：

```
21 # virsh pool-define new.xml
   Pool vm2 defined from new.xml

22 # virsh pool-list --all --details
   Name     State     Autostart   Persistent   Capacity    Allocation   Available
   ---------------------------------------------------------------------------------
   default  running   yes         yes          16.99 GiB   10.31 GiB    6.68 GiB
   iso      running   yes         yes          16.99 GiB   10.31 GiB    6.68 GiB
   vm       running   yes         yes          79.96 GiB   14.70 GiB    65.26 GiB
   vm2      inactive  no          yes          -           -            -
```

第 21 行命令会创建存储池的定义，通过第 22 行命令来检查是否成功。

接下来执行的命令如下：

```
23 # virsh pool-build vm2
   Pool vm2 built

24 # ls -ld -Z /vm2/
   drwx--x--x. 2 root root system_u:object_r:root_t:s0 6 Dec 29 10:40 /vm2/
```

第 23 行命令将构建存储池。libvirt 会检查目标路径/vm2 是否存在，如果不存在，则会创建新目录。这可以从第 24 行命令的输出中得到验证。

接下来执行的命令如下：

```
25 # virsh pool-start vm2
   Pool vm2 started
```

```
26 # mount | grep 192.168.1.233
   192.168.1.233:/vmdata on /vm2 type nfs4 (rw,relatime,vers = 4.2,rsize = 524288,wsize =
524288,namlen = 255,hard,proto = tcp,timeo = 600,retrans = 2,sec = sys,clientaddr = 192.168.1.
231,local_lock = none,addr = 192.168.1.233)
```

第 25 行命令用于启动存储池，libvirt 会将网络文件服务器的共享目录/vmdata 挂载到宿主机的目标目录/vm2 上。这可以从第 26 行命令的输出中得到验证。

接下来执行的命令如下：

```
27 # virsh pool-autostart vm2
   Pool vm2 marked as autostarted

28 # virsh pool-list --all --details
 Name     State     Autostart   Persistent   Capacity    Allocation   Available
---------------------------------------------------------------------------------
 default  running   yes         yes          16.99 GiB   10.31 GiB    6.68 GiB
 iso      running   yes         yes          16.99 GiB   10.31 GiB    6.68 GiB
 vm       running   yes         yes          79.96 GiB   14.70 GiB    65.26 GiB
 vm2      running   yes         yes          16.99 GiB   3.94 GiB     13.05 GiB
```

第 27 行命令将其设置为自动启动。第 28 行命令显示了当前存储池的列表。

与基于目录的存储池类似，基于网络文件系统的存储池也是挂载目录，其中的存储卷就是此目录中的映像文件。删除基于网络文件系统的存储池的方法也类似。

当前版本 Cockpit(cockpit-machines-211.3-1)只能使用已有的目录来作为目标路径，而不会自动创建目标路径，如图 6-18 所示。

图 6-18　在 Cockpit 中创建基于 NFS 文件系统的存储池

在 virt-manager 中的操作也类似,如图 6-19 所示。单击 XML 选项卡,可以显示新存储池的 XML 定义,如图 6-20 所示。

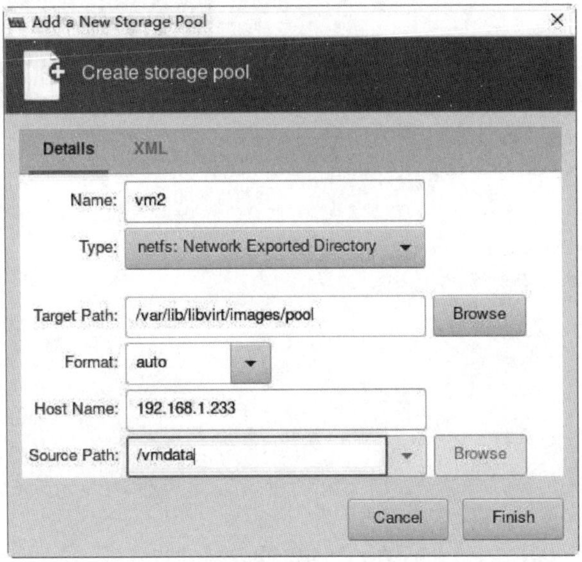

图 6-19　在 virt-manager 中创建基于网络文件系统存储池(1)

图 6-20　在 virt-manager 中创建基于网络文件系统存储池(2)

如果虚拟机的映像文件位于基于 NFS 的存储池中,而且在 SELinux 处于启动状态,则在启动虚拟机时可能出现以下错误:

```
# virsh start centos6.10
error: Failed to start domain centos6.10
error: internal error: process exited while connecting to monitor: 2021-01-05T14:36:05.
733440Z qemu-kvm: -blockdev {"driver":"file","filename":"/vm2/centos6.10.qcow2","node-
name":"libvirt-1-storage","auto-read-only":true,"discard":"unmap"}: Could not open
'/vm2/centos6.10.qcow2': Permission denied
```

如果临时禁用 SELinux 之后启动虚拟机正常,则说明还需要修改 SELinux 的布尔值,具体操作命令如下:

```
# setsebool virt_use_nfs on

# getsebool -a | grep virt_use_nfs
  virt_use_nfs --> on

# virsh start centos6.10
  Domain centos6.10 started
```

6.4.8 基于 iSCSI 目标的存储池

基于 iSCSI 目标的存储池也是一种集中的外部存储解决方案。它的存储卷位于 iSCSI 目标(Target)上的 LUN(Logical Unit Number)。当前版本的 libvirt(4.5.0)只能使用 iSCSI 目标上现有的 LUN,而不能创建新的存储卷。

在下面实验中,我们将在一台运行 CentOS 8 的主机上(主机名为 stor1,IP 地址是 192.168.1.233)安装 targetcli 软件包,通过 LinuxIO 实现 iSCSI 目标,命令如下:

```
1 [root@stor1 ~]# cat /etc/redhat-release
  CentOS Linux release 8.2.2004 (Core)

2 [root@stor1 ~]# yum -y install targetcli

3 [root@stor1 ~]# targetcli
  Warning: Could not load preferences file /root/.targetcli/prefs.bin.
  targetcli shell version 2.1.53
  Copyright 2011-2013 by Datera, Inc and others.
  For help on commands, type 'help'.
```

第 2 行命令用于安装 targetcli 软件包。

第 3 行执行 targetcli 命令,会进入操作的上下文。在此上下文中,有类似文件目录的树形结构,也有自动补全功能,还有很清晰的帮助功能。

输入 ls 命令,会显示当前的目录树。backstores 是后端存储,可以是块设备、LVM 或文件。iscsi 控制允许访问的启动器(initiator)和访问控制列表(ACL)。loopback 很少会用

到。由于是全新的环境,所以全部是空的,显示的内容如下:

```
/> ls
o- / ......................................................... [...]
  o- backstores ............................................... [...]
  | o- block ................................. [Storage Objects: 0]
  | o- fileio ................................ [Storage Objects: 0]
  | o- pscsi ................................. [Storage Objects: 0]
  | o- ramdisk ............................... [Storage Objects: 0]
  o- iscsi ............................................ [Targets: 0]
  o- loopback ......................................... [Targets: 0]
```

首先需要创建存储对象。为了简化实验操作,准备使用文件来当作后端存储。

输入 cd backstores/fileio 命令,将进入 backstores/fileio 子目录。使用 create 命令创建一个文件对象。这个对象路径名是/vmdata/disk0.img,大小为 50MB,示例命令如下:

```
/> cd backstores/fileio

/backstores/fileio> create disk0 /vmdata/disk0.img 50MB
Created fileio disk0 with size 52428800

/backstores/fileio> ls
o- fileio ..................................... [Storage Objects: 1]
  o- disk0 . [/vmdata/disk0.img (50.0MiB) write-back deactivated]
    o- alua ..................................... [ALUA Groups: 1]
      o- default_tg_pt_gp ........ [ALUA state: Active/optimized]
```

有了后端存储,就可以创建 iSCSI 目标了。使用 cd 命令进入/iscsi 目录,由于没有任何配置,所以 ls 的输出是空的,命令如下:

```
/backstores/fileio> cd /iscsi

/iscsi> ls
o- iscsi ............................................ [Targets: 0]
```

使用 create 命令创建 Target 对象。如果不指定 Target 对象的 WWN,targetcli 会自动生成一个,在本示例中是 iqn.2003-01.org.Linux-iscsi.stor1.x8664:sn.dd4c6736e8ab(以下简写为 WWN)。在创建 Target 对象的同时,还会自动创建一个名为 tpg1 的 TPG(Target Portal Group)。在 tpg1 之下,还有 ACL、LUN、Portal 等新对象,命令如下:

```
/iscsi> create
Created target iqn.2003-01.org.Linux-iscsi.stor1.x8664:sn.dd4c6736e8ab.
Created TPG 1.
Global pref auto_add_default_portal=true
```

```
Created default portal listening on all IPs (0.0.0.0), port 3260.

/iscsi> ls
o- iscsi ......................................................... [Targets: 1]
  o- iqn.2003-01.org.Linux-iscsi.stor1.x8664:sn.dd4c6736e8ab [TPGs: 1]
    o- tpg1 ............................... [no-gen-acls, no-auth]
      o- acls ................................................. [ACLs: 0]
      o- luns ................................................. [LUNs: 0]
      o- portals ............................................ [Portals: 1]
        o- 0.0.0.0:3260 ......................................... [OK]
```

进入 WWN/tpg1/luns 中，在文件型后端存储/backstores/fileio/disk0 上创建一个 LUN，命令如下：

```
/iscsi> cd iqn.2003-01.org.Linux-iscsi.stor1.x8664:sn.dd4c6736e8ab/tpg1/

/iscsi/iqn.20...6736e8ab/tpg1> luns/ create /backstores/fileio/disk0
```

再次查看 tpg1 下的信息，luns 子目录中新增了一个名称为 lun0 的资源，命令如下：

```
/iscsi/iqn.20...6736e8ab/tpg1> ls
o- tpg1 ................................... [no-gen-acls, no-auth]
  o- acls ................................................. [ACLs: 0]
  o- luns ................................................. [LUNs: 1]
  | o- lun0 [fileio/disk0 (/vmdata/disk0.img) (default_tg_pt_gp)]
  o- portals ............................................ [Portals: 1]
    o- 0.0.0.0:3260 ......................................... [OK]
```

有了 LUN 资源，就可以设置 ACL 了。ACL 需要知道 iSCSI 启动器的标识。在本实验中，iSCSI 启动器就是虚拟化的宿主机，它的标识保存在/etc/iscsi/initiatorname.iscsi 文件中。可以将随机生成的 iSCSI 启动器的名称修改为更有意义的名称，命令如下：

```
# vi /etc/iscsi/initiatorname.iscsi
# 将随机字符修改为主机名
InitiatorName=iqn.1994-05.com.redhat:kvm1
```

进入/iscsi/WWN/tgp1/acls/目录，使用 create 命令创建一条 ACL 条目，允许来自 iqn.1994-05.com.redhat:kvm1 的访问，命令如下：

```
/iscsi/iqn.20...6736e8ab/tpg1> cd acls

/iscsi/iqn.20...8ab/tpg1/acls> create iqn.1994-05.com.redhat:kvm1
```

进入根目录，使用 ls 命令查看全部配置，命令如下：

```
/iscsi/iqn.20...8ab/tpg1/acls> cd /
/> ls
o- / ......................................................................... [...]
  o- backstores .............................................................. [...]
  | o- block ................................................ [Storage Objects: 0]
  | o- fileio ............................................... [Storage Objects: 1]
  | | o- disk0 [/vmdata/disk0.img (50.0MiB) write-back activated]
  | |   o- alua ..................................................... [ALUA Groups: 1]
  | |     o- default_tg_pt_gp .... [ALUA state: Active/optimized]
  | o- pscsi ................................................. [Storage Objects: 0]
  | o- ramdisk ............................................... [Storage Objects: 0]
  o- iscsi .................................................................... [Targets: 1]
  | o- iqn.2003-01.org.Linux-iscsi.stor1.x8664:sn.dd4c6736e8ab [TPGs: 1]
  |   o- tpg1 .......................................... [no-gen-acls, no-auth]
  |     o- acls ............................................................. [ACLs: 1]
  |     | o- iqn.1994-05.com.redhat:kvm1 ....... [Mapped LUNs: 1]
  |     |   o- mapped_lun0 ............ [lun0 fileio/disk0 (rw)]
  |     o- luns ............................................................. [LUNs: 1]
  |     | o- lun0 [fileio/disk0 (/vmdata/disk0.img) (default_tg_pt_gp)]
  |     o- portals ....................................................... [Portals: 1]
  |       o- 0.0.0.0:3260 ............................................... [OK]
  o- loopback ............................................................. [Targets: 0]
```

检查无误后使用 saveconfig 命令保存配置，然后使用 exit 命令退出 targetcli 上下文，命令如下：

```
/> saveconfig
Configuration saved to /etc/target/saveconfig.json

/> exit
Global pref auto_save_on_exit=true
Last 10 configs saved in /etc/target/backup/.
Configuration saved to /etc/target/saveconfig.json
```

配置存储服务器的防火墙，允许 iSCSI 协议（TCP 3260）的入站，命令如下：

```
[root@stor1 ~]# firewall-cmd --permanent --add-service=iscsi-target

[root@stor1 ~]# firewall-cmd --reload
```

将 target 服务设置为自动启动，然后重新启动 target 服务，命令如下：

```
[root@stor1 ~]# systemctl enable target.service

[root@stor1 ~]# systemctl restart target
```

至此,实验用的存储服务器配置完毕。下面我们在虚拟化的宿主机上创建基于 iSCSI 目标的存储池。

在创建存储池之前,要保证 iSCSI 启动器(也就是虚拟化的宿主机)可以正确地连接 iSCSI 目标,命令如下:

```
1 # iscsiadm -- mode discovery -- type sendtargets -- portal 192.168.1.233
  192.168.1.233:3260,1 iqn.2003-01.org.Linux-iscsi.stor1.x8664:sn.dd4c6736e8ab

2 # iscsiadm -d2 -m node -- login
  iscsiadm: Max file limits 1024 262144
  iscsiadm: default: Creating session 1/1
  Logging in to [iface: default, target: iqn.2003-01.org.Linux-iscsi.stor1.x8664:sn.dd4c6736e8ab, portal: 192.168.1.233,3260]
  Login to [iface: default, target: iqn.2003-01.org.Linux-iscsi.stor1.x8664:sn.dd4c6736e8ab, portal: 192.168.1.233,3260] successful.
```

第 1 行命令用于检查是否可以发现 iSCSI 目标。

第 2 行命令尝试登录。

接下来执行的命令如下:

```
3 # fdisk -l
  ...
  Disk /dev/sdc: 50 MiB, 52428800 Bytes, 102400 sectors
  Units: sectors of 1 * 512 = 512 Bytes
  Sector size (logical/physical): 512 Bytes / 512 Bytes
  I/O size (minimum/optimal): 512 Bytes / 8388608 Bytes

4 # iscsiadm -d2 -m node -- logout
  iscsiadm: Max file limits 1024 262144
  Logging out of session [sid: 1, target: iqn.2003-01.org.Linux-iscsi.stor1.x8664:sn.dd4c6736e8ab, portal: 192.168.1.233,3260]
  Logout of [sid: 1, target: iqn.2003-01.org.Linux-iscsi.stor1.x8664:sn.dd4c6736e8ab, portal: 192.168.1.233,3260] successful.
```

由于存储服务器配置了允许的 ACL,所以在第 3 行命令的输出中会看到新的磁盘 /dev/sdc,它就是 iSCSI 目标上的 LUN0。

这说明测试成功,可以使用第 4 行命令断开与 Target 的连接。

在 iSCSI 启动器与 iSCSI 目标连通性测试通过之后,就可以创建新的存储池了。首先需要创建一个 XML 文件,用于保存新存储池的参数。参数的含义如表 6-7 所示,命令如下:

```
5 # vi new.xml
  <pool type="iscsi">
    <name>vm2</name>
    <source>
      <host name="192.168.1.233"/>
      <device path="iqn.2003-01.org.Linux-iscsi.stor1.x8664:sn.dd4c6736e8ab"/>
    </source>
    <target>
      <path>/dev</path>
    </target>
  </pool>
```

表 6-7 基于 iSCSI 目标存储池的参数

描述	XML 格式
存储池的类型	`<pool type='iscsi'>`
存储池的名称	`<name>name</name>`
存储服务器的主机名或 IP 地址	`<source><host name='hostname' />`
iSCSI IQN	`<device path="iSCSI_IQN" /></source>`
存储池的目标路径	`<target>`
	`<path>/dev</path>`
	`</target>`

接下来执行的命令如下：

```
6 # virsh pool-define new.xml
  Pool vm2 defined from new.xml

7 # virsh pool-list --all --details
  Name       State      Autostart   Persistent   Capacity    Allocation   Available
  -------------------------------------------------------------------------------------
  default    running    yes         yes          16.99 GiB   10.30 GiB    6.68 GiB
  iso        running    yes         yes          16.99 GiB   10.30 GiB    6.68 GiB
  vm         running    yes         yes          79.96 GiB   14.70 GiB    65.26 GiB
  vm2        inactive   no          yes
```

第 6 行命令会创建存储池的定义。

从第 7 行命令的输出可以看出：新的存储池处于非活动状态。

接下来执行的命令如下：

```
8 # virsh pool-build vm2

9 # virsh pool-start vm2
```

```
10 #fdisk -l | grep "^Disk /dev"
   Disk /dev/sdb: 80 GiB, 85899345920 Bytes, 167772160 sectors
   Disk /dev/sda: 20 GiB, 21474836480 Bytes, 41943040 sectors
   Disk /dev/mapper/cl-root: 17 GiB, 18249416704 Bytes, 35643392 sectors
   Disk /dev/mapper/cl-swap: 2 GiB, 2147483648 Bytes, 4194304 sectors
   Disk /dev/mapper/vgvm1-lvvm1: 80 GiB, 85895151616 Bytes, 167763968 sectors
   Disk /dev/sdc: 50 MiB, 52428800 Bytes, 102400 sectors
```

第 8 行命令用于构建存储池。

第 9 行命令用于启动存储池。libvirt 会使用 iSCSI 启动器发起到 iSCSI 目标的连接，从而将其上的资源 LUN0 附加到 /dev 目录下，这可以从第 10 行命令的输出得到验证：有新的块设备 /dev/sdc。

接下来执行的命令如下：

```
11 #virsh pool-dumpxml vm2
   <pool type='iscsi'>
     <name>vm2</name>
     <uuid>14072507-6601-4329-832c-39d7ebf8ebf9</uuid>
     <capacity unit='Bytes'>73400320</capacity>
     <allocation unit='Bytes'>73400320</allocation>
     <available unit='Bytes'>0</available>
     <source>
       <host name='192.168.1.233' port='3260'/>
       <device path='iqn.2003-01.org.Linux-iscsi.stor1.x8664:sn.dd4c6736e8ab'/>
     </source>
     <target>
       <path>/dev</path>
     </target>
   </pool>
```

第 11 行命令用于查看新存储池的详细信息。

接下来执行的命令如下：

```
12 #virsh vol-list --pool vm2 --details
    Name          Path        Type    Capacity    Allocation
   -----------------------------------------------------------
    unit:0:0:0    /dev/sdc    block   50.00 MiB   50.00 MiB

13 #virsh vol-create-as vm2 test1 10M
   error: Failed to create vol test1
   error: this function is not supported by the connection driver: storage pool does not support volume creation
```

从第 12 行命令的输出可以看出：iSCSI 目标上的 LUN 资源就是存储池中的存储卷。

第 13 行命令输出错误，这是因为 libvirt 是不能在基于 iSCSI 目标的存储池中创建新的存储卷的。要想创建存储卷，还需要在 iSCSI 目标上进行操作。

使用 Cockpit 创建基于 iSCSI 目标的存储池，如图 6-21 所示。

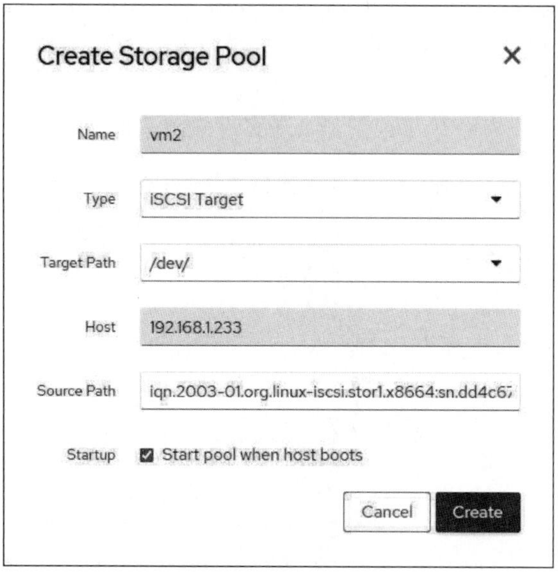

图 6-21　在 Cockpit 中创建基于 iSCSI 目标的存储池

在 virt-manager 中的操作也类似，如图 6-22(a) 所示。单击 XML 选项卡，可以显示新存储池的 XML 定义，如图 6-22(b) 所示。

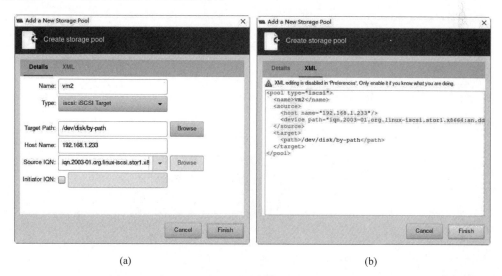

图 6-22　在 virt-manager 中创建基于 iSCSI 目标的存储池

6.5 存储卷

可以将存储池划分为存储卷(Storage Volume)。存储卷可以是映像文件、物理分区、LVM 逻辑卷或其他可以被 libvirt 管理的存储。无论底层硬件架构如何,存储卷都将作为块存储设备呈现给虚拟机。

virsh 有多个以 vol 开头的并与存储卷有关的子命令,如表 6-8 所示。

表 6-8 virsh 中存储卷管理子命令

virsh 子命令	功能说明
vol-clone	根据现有卷克隆出新卷
vol-create-as	根据参数来创建新卷
vol-create	根据 XML 文件创建新卷
vol-create-from	使用另外一个卷作为输入创建新卷
vol-delete	删除卷
vol-download	将卷中内容下载到文件
vol-dumpxml	以 XML 格式显示卷的详细信息
vol-info	显示卷的基本信息
vol-key	根据卷名或路径返回卷的键值
vol-list	显示存储中卷的列表
vol-name	根据卷键值或路径返回卷的名称
vol-path	根据卷名或键值返回卷的路径
vol-pool	根据卷键值或路径的返回卷
vol-resize	调整卷的大小
vol-upload	将文件内容上传到卷
vol-wipe	擦除卷中的数据

6.5.1 获得存储卷的信息

执行的命令如下:

```
1 # virsh vol-list vm --details
 Name                Path                    Type    Capacity    Allocation
---------------------------------------------------------------------------
 centos6.10.qcow2    /vm/centos6.10.qcow2    file    10.00 GiB   1.08 GiB
 disk1.img           /vm/disk1.img           file    1.00 GiB    1.00 GiB
 disk2-seek.img      /vm/disk2-seek.img      file    11.00 GiB   1.00 GiB
 win2k19.qcow2       /vm/win2k19.qcow2       file    50.00 GiB   9.23 GiB
 win2k3.qcow2        /vm/win2k3.qcow2        file    20.00 GiB   1.80 GiB
```

第 1 行命令输出了存储池 vm2 中的存储卷的列表。使用可选的--details 选项,会额外显示卷类型、总容量、已分配容量等信息。

接下来执行的命令如下：

```
2 # virsh vol-info --pool vm --vol centos6.10.qcow2
  Name:           centos6.10.qcow2
  Type:           file
  Capacity:       10.00 GiB
  Allocation:     1.08 GiB

3 # virsh vol-key --pool vm --vol centos6.10.qcow2
  /vm/centos6.10.qcow2

4 # virsh vol-path --pool vm --vol centos6.10.qcow2
  /vm/centos6.10.qcow2
```

第 2 行命令显示存储池 vm 中的一个名为 centos6.10.qcow2 的存储卷的基本信息。libvirt 可以使用 3 种方法来标识特定的存储卷。

(1) 存储池名称＋存储卷名称：第 2 行命令就是采用的这种方法，存储池 vm 中的存储卷为 centos6.10.qcow2。

(2) 存储卷的键值(key)：对于基于目录的存储池中的映像文件来讲，它的键值就是文件的路径名。第 3 行命令用于将存储池名称＋存储卷名称转换为键值。

(3) 存储卷的路径：第 4 行命令将存储池名称＋存储卷名称转换为路径。

接下来执行的命令如下：

```
5 # virsh vol-dumpxml --pool vm --vol centos6.10.qcow2
  <volume type='file'>
    <name>centos6.10.qcow2</name>
    <key>/vm/centos6.10.qcow2</key>
    <source>
    </source>
    <capacity unit='Bytes'>10737418240</capacity>
    <allocation unit='Bytes'>1157107712</allocation>
    <physical unit='Bytes'>1157169152</physical>
    <target>
      <path>/vm/centos6.10.qcow2</path>
      <format type='qcow2'/>
      <permissions>
        <mode>0644</mode>
        <owner>0</owner>
        <group>0</group>
        <label>system_u:object_r:default_t:s0</label>
      </permissions>
      <timestamps>
```

```
            <atime>1609121006.309430030</atime>
            <mtime>1608282428.678574373</mtime>
            <ctime>1608282440.051573949</ctime>
        </timestamps>
        <compat>1.1</compat>
        <features/>
    </target>
</volume>
```

第 5 行命令输出了存储卷的详细信息。

接下来执行的命令如下：

```
6 # virsh vol-pool /vm/centos6.10.qcow2
  vm

7 # virsh vol-pool /vm/centos6.10.qcow2 --uuid
  13c17fca-c020-4ca2-9ea0-8cbe07a62d23
```

还可以使用 vol-pool 子命令根据存储卷的路径名来查找其所归属的存储池。第 6 行显示 /vm/centos6.10.qcow2 属于存储池 vm。如果像第 7 行一样再加上 --uuid 选项，则会输出存储池的 UUID。

6.5.2 创建存储卷

由于存储池已经将底层的存储硬件进行了抽象，所以在不同类型存储池中创建卷的操作基本相同。根据创建新存储卷参数的由来划分，virsh 有 4 种创建存储卷的方法。

(1) 根据参数创建新卷：vol-create-as。
(2) 根据 XML 文件创建新卷：vol-create。
(3) 根据现有卷创建新卷：vol-create-from。
(4) 根据现有卷克隆出新卷：vol-clone。

1. 根据参数创建新卷

子命令 vol-create-as 可以根据命令行参数创建新卷。其语法格式如下：

```
# virsh vol-create-as <pool> <name> <capacity> [--allocation <string>] [--format
<string>] [--backing-vol <string>] [--backing-vol-format <string>] [--prealloc-
metadata] [--print-xml]
```

(1) [--pool] string：必需参数，指定存储池的名称。

(2) [--name] string：必需参数，指定新存储卷的名称。

(3) [--capacity] string：必需参数，指定新存储卷的大小，以整数表示。可以使用的后缀有 b、k、M、G、T，分别表示字节、千字节、兆字节、千兆字节和太字节。默认为字节。

（4）--allocation string：指定初始分配大小，以整数表示。默认为字节。

（5）--format string：指定文件格式类型。仅适用于基于文件的存储池，可接受的类型包括 raw、bochs、qcow、qcow2、qed、host_device 和 vmdk。默认格式是 raw。

（6）--backing-vol string：指定基础卷。创建快照时，会使用此选项。

（7）--backing-vol-format string：指定基础卷的格式。创建快照时，会使用此选项。

（8）--prealloc-metadata：预分配元数据。

根据参数创建新卷的命令如下：

```
1 # virsh vol-create-as --pool vm2 --name test1.qcow2 --capacity 500M --format qcow2
  Vol test1.qcow2 created

2 # virsh vol-list vm2 --details
  Name         Path                Type    Capacity      Allocation
  ---------------------------------------------------------------
  test1.qcow2  /vm2/test1.qcow2    file    500.00 MiB    196.00 KiB

3 # qemu-img info /vm2/test1.qcow2
  image: /vm2/test1.qcow2
  file format: qcow2
  virtual size: 500M (524288000 Bytes)
  disk size: 196K
  cluster_size: 65536
  Format specific information:
      compat: 0.10
      refcount bits: 16

4 # ls -lZ /vm2
  total 196
  -rw-------. 1 root root system_u:object_r:root_t:s0 196616 Dec 27 11:46 test1.qcow2
```

第 1 行命令在存储池 vm2 中创建了一个新卷 test1.qcow2，其容量为 500MiB，格式为 qcow2。

第 2 行、第 3 行命令显示了新存储卷的信息，它的格式是精简供给。

第 4 行命令显示文件的 SELinux 的上下文为 system_u:object_r:root_t:s0。

2．根据 XML 文件创建新卷

子命令 vol-create 根据包含存储卷参数的 XML 文件创建一个新的存储卷。其语法格式如下：

```
# virsh vol-create <pool> <file> [--prealloc-metadata]
```

在一个基于目录的存储池中创建新的存储卷，命令如下：

```
1 # vi new.xml
  <volume>
    <name>test1.qcow2</name>
    <capacity>1073741824</capacity>
    <allocation>0</allocation>
    <target>
      <format type="qcow2"/>
    </target>
  </volume>

2 # virsh vol-create vm2 new.xml
  Vol test1.qcow2 created from new.xml
```

第1行命令创建了一个新的 XML 文件,它保存了新存储卷的参数。新卷的名称为 test1.qcow2,容量是 1GB,格式为 qcow2。

第2行命令用于创建新存储卷。

接下来执行的命令如下:

```
3 # virsh vol-list vm2 --details
   Name          Path              Type    Capacity    Allocation
  ------------------------------------------------------------------
   test1.qcow2   /vm2/test1.qcow2  file    1.00 GiB    196.00 KiB

4 # qemu-img info /vm2/test1.qcow2
  image: /vm2/test1.qcow2
  file format: qcow2
  virtual size: 1.0G (1073741824 Bytes)
  disk size: 196K
  cluster_size: 65536
  Format specific information:
      compat: 0.10
      refcount bits: 16
```

第3行、第4行命令显示了新存储卷的信息。映像文件的路径为/vm2/test1.qcow2。

3. 根据现有卷创建新卷

子命令 vol-create-from 与 vol-create 很类似,用于创建新的存储卷。不过它在创建新卷时会以一个现有卷作为输入,这样新卷具有与现有卷相同的数据。其语法格式如下:

```
# virsh vol-create-from <pool> <file> <vol> [--inputpool <string>] [--prealloc-metadata] [--reflink]
```

(1)--pool string:必需参数,指定新存储卷所在的存储池的名称或 UUID。该存储池不必与现有的存储卷的存储池相同。

（2）--file string：必需参数，指定包含存储卷的参数的 XML 文件。
（3）--vol string：必需参数，指定现有存储卷的名称或路径。
（4）--inputpool string：可选参数，指定现有存储卷所在的存储池名称。
（5）--prealloc-metadata：可选参数，设置预分配元数据。

根据现有卷创建新卷的命令如下：

```
1 # cat new.xml
  <volume>
    <name>test2.qcow2</name>
    <capacity>1073741824</capacity>
    <allocation>0</allocation>
    <target>
      <format type="qcow2"/>
    </target>
  </volume>

2 # virsh vol-create-from --pool vm --file new.xml /vm2/test1.qcow2
  Vol test2.qcow2 created from input vol test1.qcow2

3 # virsh vol-info --pool vm --vol test2.qcow2
  Name:        test2.qcow2
  Type:        file
  Capacity:    1.00 GiB
  Allocation:  1.25 MiB
```

第 1 行命令创建了新卷的定义。

第 2 行命令会在存储池 vm 中创建一个名为 test2.qcow2 的新卷，新卷中的数据来自 /vm2/test1.qcow2。

第 3 行命令显示了新存储卷的信息。

4. 根据现有卷克隆出新卷

子命令 vol-clone 是 vol-create-from 的一个简化的、易用的版本。它会在相同的池中创建一个现有卷的副本。其语法格式如下：

```
# virsh vol-clone <vol> <newname> [--pool <string>] [--prealloc-metadata] [--reflink]
```

由于是克隆，所以无须为新卷指定大小、格式等属性。下面把存储池 vm2 的存储卷 test1.qcow2 克隆一份，新卷名称为 test2.qcow2，命令如下：

```
1 # virsh vol-list vm2
   Name              Path
  ------------------------------------------------------------
   test1.qcow2       /vm2/test1.qcow2
```

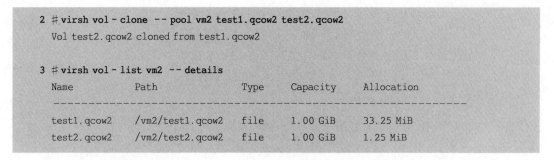

在 Cockpit 中创建存储卷时，需要为新卷设置名称及大小。

如果在基于目录的存储池中创建存储卷，则还需要设置映像文件格式，如图 6-23 所示。

如果在基于物理磁盘的存储池中创建存储卷，则新卷的名称需要符合分区命名规范（例如：sdc1），如图 6-24 所示。

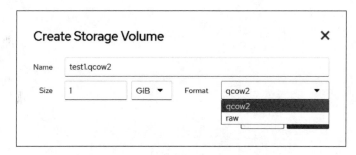

图 6-23　在基于目录的存储池中创建存储卷

图 6-24　在基于物理磁盘的存储池中创建存储卷

与 Cockpit 相比，在 virt-manager 中创建新卷时，virt-manager 的选项更多一些，例如可以通过 Backing store 设置基础映像文件，如图 6-25 和图 6-26 所示。

6.5.3　向虚拟机分配存储卷

virsh 有 3 个子命令可以用来向虚拟机分配存储卷。

（1）edit：通过编辑虚拟机的 XML 文件来管理存储。这种方法工作量大、易出错。

图 6-25　在基于目录的存储池中创建存储卷

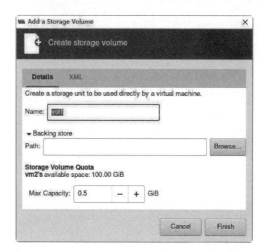

图 6-26　在基于 LVM 中的存储池中创建存储卷

（2）attach-device：通过 XML 文件向虚拟机添加包括存储设备在内的新设备。

（3）attach-disk：通过参数向虚拟机添加新的存储设备，可以认为它是 attach-device 的一种再封装的简化版本。

对于 libvirt 当前版本（4.5.0），attach-disk 子命令还只能通过存储卷的路径名（例如：/vm/centos6.10.qcow2）来指定存储卷，而无法使用"存储池名称＋存储卷名称"的方式来指定存储卷。virt-manager 与其类似，而 attach-device 子命令和 Cockpit 则可以使用"存储池名称＋存储卷名称"的方式，所以可以充分利用存储池在存储管理的灵活性。

使用 attach-device 子命令的示例命令如下：

```
1 # virsh attach-device --help
  NAME
    attach-device - attach device from an XML file

  SYNOPSIS
    attach-device <domain> <file> [--persistent] [--config] [--live] [--current]

  DESCRIPTION
    Attach device from an XML <file>.

  OPTIONS
    [--domain] <string>  domain name, id or uuid
    [--file] <string>    XML file
    --persistent         make live change persistent
    --config             affect next boot
    --live               affect running domain
    --current            affect current domain
```

主要参数说明如下。

(1) --domain：指定虚拟机的名称（或 ID、UUID）。

(2) --file：指定 XML 文件。

(3) --config：指定下次启动后生效。本次修改的是配置文件。

(4) --live：修改正在运行的虚拟机。

(5) --current：如果虚拟机正在运行，就修改当前的，相当于--live。如果虚拟机没有运行，就修改配置文件，相当于--config。

接下来执行的命令如下：

```
2 # vi new.xml
  <disk type='volume' device='disk'>
    <driver name='qemu' type='qcow2'/>
    <source pool='vm2' volume='test1.qcow2'/>
    <target dev='vdb' bus='virtio'/>
  </disk>

3 # virsh attach-device centos6.10 new.xml --config
  Device attached successfully

4 # virsh domblklist centos6.10
  Target     Source
  ------------------------------------------------
  vda        /var/lib/libvirt/images/centos6.10.qcow2
  vdb        test1.qcow2
  hda        -
```

第 2 行命令用于创建 XML 文件。通过 <source pool='vm2' volume='test1.qcow2'/> 指定存储池和存储卷。

第 3 行命令根据 XML 向虚拟机添加新磁盘。

从第 4 行命令的输出可以看出，vda 使用的是存储卷路径名，而 vdb 使用的是"存储池名称＋存储卷名称"。

提示：attach-disk 子命令的使用，参见 6.2.2 使用 virsh 管理虚拟机磁盘映像文件一节。

6.5.4 删除存储卷及擦除存储卷

vol-delete 子命令会从存储池中删除存储卷信息。根据存储池类型的不同，删除操作的结果会有所差异，例如：基于目录的存储池是删除映像文件，基于物理磁盘的存储池则是删除分区，而基于 LVM 的存储池是删除逻辑卷。

删除基于目录的存储池中的存储卷，命令如下：

```
1 #virsh vol-list vm2
   Name                    Path
  -----------------------------------------------------------
   test1.qcow2             /vm2/test1.qcow2
   test2.qcow2             /vm2/test2.qcow2

2 #virsh vol-delete --pool vm2 --vol test1.qcow2
   Vol test1.qcow2 deleted
```

第 1 行命令显示了存储池 vm2 的存储卷的列表。

第 2 行命令用于删除指定卷。还可以通过卷的键值或路径指定存储卷。

使用 vol-delete 子命令删除存储卷是一种"逻辑"删除，有可能通过一些恢复工具来恢复部分或全部数据，所以建议在生产环境中先擦除数据再进行删除。

使用 vol-wipe 子命令可以擦除卷中的数据，这样可以避免恢复先前的数据。在使用时，可以通过 --algorithm 选项来指定擦除算法，包括 zero、nnsa、dod、bsi、gutmann、schneier、pfitzner7、pfitzner33、random 等。

擦除卷中的数据的命令如下：

```
3 #virsh vol-wipe --pool vm2 --vol test2.qcow2
   Vol test2.qcow2 wiped

4 #virsh vol-delete --pool vm2 --vol test2.qcow2
   Vol test2.qcow2 deleted
```

第 3 行命令先擦除存储卷的数据,然后执行第 4 行命令进行删除。

6.6 本章小结

本章讲解了虚拟存储的知识,包括托管和非托管的存储区别,qemu-img 命令的使用,以及存储池、存储卷的原理等。

图 书 推 荐

书 名	作 者
仓颉语言实战（微课视频版）	张磊
仓颉语言元编程	张磊
仓颉语言核心编程——入门、进阶与实战	徐礼文
仓颉语言程序设计	董昱
仓颉程序设计语言	刘安战
仓颉语言极速入门——UI全场景实战	张云波
HarmonyOS 移动应用开发（ArkTS版）	刘安战、余雨萍、陈争艳 等
openEuler 操作系统管理入门	陈争艳、刘安战、贾玉祥 等
深度探索 Vue.js——原理剖析与实战应用	张云鹏
前端三剑客——HTML5＋CSS3＋JavaScript 从入门到实战	贾志杰
剑指大前端全栈工程师	贾志杰、史广、赵东彦
Flink 原理深入与编程实战——Scala＋Java（微课视频版）	辛立伟
Spark 原理深入与编程实战（微课视频版）	辛立伟、张帆、张会娟
PySpark 原理深入与编程实战（微课视频版）	辛立伟、辛雨桐
HarmonyOS 应用开发实战（JavaScript 版）	徐礼文
HarmonyOS 原子化服务卡片原理与实战	李洋
鸿蒙操作系统开发入门经典	徐礼文
鸿蒙应用程序开发	董昱
鸿蒙操作系统应用开发实践	陈美汝、郑森文、武延军、吴敬征
HarmonyOS App 开发从 0 到 1	张诏添、李凯杰
JavaScript 修炼之路	张云鹏、戚爱斌
JavaScript 基础语法详解	张旭乾
华为方舟编译器之美——基于开源代码的架构分析与实现	史宁宁
Android Runtime 源码解析	史宁宁
恶意代码逆向分析基础详解	刘晓阳
网络攻防中的匿名链路设计与实现	杨昌家
深度探索 Go 语言——对象模型与 runtime 的原理、特性及应用	封幼林
深入理解 Go 语言	刘丹冰
Vue＋Spring Boot 前后端分离开发实战	贾志杰
Spring Boot 3.0 开发实战	李西明、陈立为
Flutter 组件精讲与实战	赵龙
Flutter 组件详解与实战	［加］王浩然（Bradley Wang）
Dart 语言实战——基于 Flutter 框架的程序开发（第 2 版）	亢少军
Dart 语言实战——基于 Angular 框架的 Web 开发	刘仕文
IntelliJ IDEA 软件开发与应用	乔国辉
Python 量化交易实战——使用 vn.py 构建交易系统	欧阳鹏程
Python 从入门到全栈开发	钱超
Python 全栈开发——基础入门	夏正东
Python 全栈开发——高阶编程	夏正东
Python 全栈开发——数据分析	夏正东
Python 编程与科学计算（微课视频版）	李志远、黄化人、姚明菊 等

续表

书　名	作　者
Diffusion AI 绘图模型构造与训练实战	李福林
HuggingFace 自然语言处理详解——基于 BERT 中文模型的任务实战	李福林
图像识别——深度学习模型理论与实战	于浩文
数字 IC 设计入门（微课视频版）	白栎旸
动手学推荐系统——基于 PyTorch 的算法实现（微课视频版）	於方仁
人工智能算法——原理、技巧及应用	韩龙、张娜、汝洪芳
Python 数据分析实战——从 Excel 轻松入门 Pandas	曾贤志
Python 概率统计	李爽
Python 数据分析从 0 到 1	邓立文、俞心宇、牛瑶
从数据科学看懂数字化转型——数据如何改变世界	刘通
鲲鹏架构入门与实战	张磊
鲲鹏开发套件应用快速入门	张磊
华为 HCIA 路由与交换技术实战	江礼教
华为 HCIP 路由与交换技术实战	江礼教
精讲 MySQL 复杂查询	张方兴
5G 核心网原理与实践	易飞、何宇、刘子琦
Python 游戏编程项目开发实战	李志远
编程改变生活——用 Python 提升你的能力（基础篇·微课视频版）	邢世通
编程改变生活——用 Python 提升你的能力（进阶篇·微课视频版）	邢世通
编程改变生活——用 PySide6/PyQt6 创建 GUI 程序（基础篇·微课视频版）	邢世通
编程改变生活——用 PySide6/PyQt6 创建 GUI 程序（进阶篇·微课视频版）	邢世通
FFmpeg 入门详解——音视频原理及应用	梅会东
FFmpeg 入门详解——SDK 二次开发与直播美颜原理及应用	梅会东
FFmpeg 入门详解——流媒体直播原理及应用	梅会东
FFmpeg 入门详解——命令行与音视频特效原理及应用	梅会东
FFmpeg 入门详解——音视频流媒体播放器原理及应用	梅会东
精讲 MySQL 复杂查询	张方兴
Python Web 数据分析可视化——基于 Django 框架的开发实战	韩伟、赵盼
Python 玩转数学问题——轻松学习 NumPy、SciPy 和 Matplotlib	张骞
Pandas 通关实战	黄福星
深入浅出 Power Query M 语言	黄福星
深入浅出 DAX——Excel Power Pivot 和 Power BI 高效数据分析	黄福星
从 Excel 到 Python 数据分析：Pandas、xlwings、openpyxl、Matplotlib 的交互与应用	黄福星
云原生开发实践	高尚衡
云计算管理配置与实战	杨昌家
虚拟化 KVM 极速入门	陈涛
虚拟化 KVM 进阶实践	陈涛
HarmonyOS 从入门到精通 40 例	戈帅
OpenHarmony 轻量系统从入门到精通 50 例	戈帅
AR Foundation 增强现实开发实战（ARKit 版）	汪祥春
AR Foundation 增强现实开发实战（ARCore 版）	汪祥春